T0220943

Computational and Statistical Methods for Protein Quantification by Mass Spectrometry

Computational and Statistical Methods for Protein Quantification by Mass Spectrometry

Ingvar Eidhammer

Department of Informatics, University of Bergen, Norway

Harald Barsnes

Department of Biomedicine, University of Bergen, Norway

Geir Egil Eide

*Centre for Clinical Research, Haukeland University
Hospital and Department of Public Health and Primary Health Care,
University of Bergen, Norway*

Lennart Martens

*Department of Biochemistry, Faculty of Medicine and
Health Sciences, Ghent University, Belgium*

A John Wiley & Sons, Ltd., Publication

This edition first published 2013
© 2013 John Wiley & Sons, Ltd

Registered office
John Wiley & Sons Ltd, The Atrium, Southern Gate, Chichester, West Sussex, PO19 8SQ, United
Kingdom

For details of our global editorial offices, for customer services and for information about how to apply
for permission to reuse the copyright material in this book please see our website at www.wiley.com.

Library of Congress Cataloging-in-Publication Data applied for.

A catalogue record for this book is available from the British Library.

ISBN: 978-1-119-96400-1

Typeset in 10/12pt Times by Aptara Inc., New Delhi, India

Contents

Preface

Mass spectrometry based proteomics has quickly become the method of choice for the high-throughput analysis of entire proteomes. Through the advent of a variety of powerful approaches for quantitative proteomics, these experiments now also yield large volumes of quantitative data on protein expression levels.

The ability to quantify a proteome in high-throughput opens up many new possibilities for research, including protein biomarker discovery, fundamental insight into the dynamics of a cell or tissue proteome over time, and analysis of expression changes in function of (external) perturbations. Such analyses in turn benefit both biological insight as well as clinical applications, and form a crucial element of the emerging field of systems biology.

Adding quantitative data to the already complex output of a proteomics experimentation does however further increase the importance of correct data processing and results evaluation. As such, the computational methods employed to analyze such data are increasingly seen as a crucial step in the overall workflow.

With this in mind the intention of this book is to systemize and describe the different experimental approaches for protein quantification by mass spectrometry, and to present the corresponding computational and statistical methods used to analyze data from such experiments.

The first three chapters act as an introduction, introducing terms, concepts, and protein quantification as well as its relation to mRNA quantification. Chapter 4 then provides a brief introduction to mass spectrometry based proteomics, of which a more detailed description can be found in Eidhammer et al. (2007). Chapter 5 systematizes and classifies the different quantification approaches that are described in more detail in Chapters 9 to 15, each of which are dedicated to a specific approach. Statistical data processing and analysis for quantitative proteomics experiments form the subject of Chapters 6, 7, and 8. The final four Chapters (16 to 19) deal with more specific quantification tasks and a few orthogonal topics associated with protein quantification.

Given that statistics form an essential part of the computational methods used to process, analyze, and interpret quantitative mass spectrometry data, we have also included two appendices that can be used as a reference on the statistical knowledge required for a complete understanding of the methods described in the various chapters.

It is not the goal of this book to provide an exhaustive theoretical foundation for the computational analysis of quantitative proteomics, but rather to present the main challenges, and extract and systematize the common principles used in solving these problems. The presentation is illustrated throughout by numerous figures and examples.

We have tried to restrict the description of a given subject to one section or chapter, but for some of the subjects it is unavoidable to treat them several times in different contexts. In these cases we have included the necessary cross references.

Note that we have also included many references and websites, although we are aware that websites can change rapidly.

The book is directed at biologists, (bio-)informaticians, and statisticians alike, and is aimed at readers across the academic spectrum, from advanced undergraduate students to post docs first entering the field of protein quantification.

Terminology

Many of the terms used in molecular biology and proteomics do not have a unique or commonly accepted definition. We have mainly tried to follow the IUPAC (International Union of Pure and Applied Chemistry) Compendium of analytical nomenclature. They have an ongoing revision of the terms used in the larger field of mass spectrometry, and a first draft exists.

IUPAC   http://www.iupac.org/publications/analytical_compendium
Ongoing revision   http://www.msterms.com/
First draft   http://www.sgms.ch/links/IUPAC_MS_Terms_Draft.pdf

We have endeavored to employ customarily used terms wherever possible, defining our own terms only where it was necessary or more appropriate for clarity. When several synonyms exist for a term, these are also given.

Acknowledgements

Numerous papers and websites have been used in the writing of this book, and citations have been given for these works. However, in order to enhance readability we have tried to minimize literature references in the text, and have instead included a bibliographic section at the end of each chapter. The sources we consulted are detailed there, as well as additional literature that may be interesting or relevant to the reader. It is our sincere hope that this structure will serve to provide the authors of the source literature with the appropriate acknowledgments.

In addition, there are a lot of people whom we have consulted during the work, and we especially want to thank Rein Aasland, Thin Thin Aye, Frode Berven, Niklaas Colaert, Sven Degroeve, Olav Mjaavatten, Jill Anette Opsahl, Eystein Oveland, Kjell Petersen, Pål Puntervoll, and An Staes for helpful discussions and for sharing their insights and profound knowledge.

IE dedicates his work to his wife Maria.

GEE thanks the Centre for Clinical Research, Haukeland University Hospital and the Departement of Public Health and Primary Health Care, University of Bergen for approving and providing facilities to work with this book. Also he thanks his colleagues in the Western Regional Health Authorities Statisticians Network and in the Life Style Epidemiology Research Group for inspiring work environment, as well as his co-workers in numerous medical and health related research projects through the years. Finally, GEE sincerely thanks his wife Kirsti, his family and his friends for continuing love and support.

HB would like to thank his friends and colleagues at the Proteomics Unit at the Department of Biomedicine, and at the Department of Informatics / the Computational Biology Unit for numerous interesting discussions and collaborations over the last years. Thanks also to my international collaborators in the extended Computational Omics and Systems Biology group and the PRIDE team at the European Bioinformatics Institute. Our ongoing collaborations make it a lot easier to see how the work I do every day constitutes a small part of the bigger puzzle that is proteomics. Finally, I would like to thank my friends and family for their continuing support.

LM would like to thank Ghent University, VIB, and his Computational Omics and Systems Biology (CompOmics) group members for creating a vibrant and intellectually stimulating atmosphere, and for many useful discussions. Special thanks go out to my wife Leen, and two sons, Ruben and Alexander, for their patience and understanding during the long hours of writing. Ruben, at 4 years of age, was fond of asking why daddy was always working on his computer, while Alexander at 9 months was simply looking bemused at the incessant sound of fingers tapping away at the keyboard. For better or for worse, both of them will now have to learn to live with a lot more fatherly attention!

1

Introduction

Numerous regulatory processes in an organism occur by changing the amount of a specific protein or a group of proteins, or by modifying proteins, and thereby making several variants of the same protein. Understanding these diverse processes is essential for progress in a long list of research areas, including biotechnology, biomedicine, and toxicology.

Given that proteins are the executive molecules in the organism, it follows that knowing which proteins are synthesized in which cells and under which conditions is vital. In other words, the identity of the proteins does not tell the whole story, of equal importance is the amount of proteins synthesized and occurring in the cells at a given time. Hence to understand the function and regulatory mechanisms of an organism, performing *protein quantification* is essential.

Before going into the details of protein quantification it is helpful to have a simple model of an organism, and knowledge about the most commonly used concepts.

1.1 The composition of an organism

An organism is a living object that can react to stimuli, grow, reproduce, and is capable of maintaining stability (homeostasis). Organism is used both for denoting individuals and a collection of individuals. Examples are viruses, bacteria, plants, animals, and individuals from these.

1.1.1 A simple model of an organism

A simple model of an organism includes the concepts *cell*, *tissue*, and *organ*, as shown in Figure 1.1.

Computational and Statistical Methods for Protein Quantification by Mass Spectrometry,
First Edition. Ingvar Eidhammer, Harald Barsnes, Geir Egil Eide and Lennart Martens.
© 2013 John Wiley & Sons, Ltd. Published 2013 by John Wiley & Sons, Ltd.

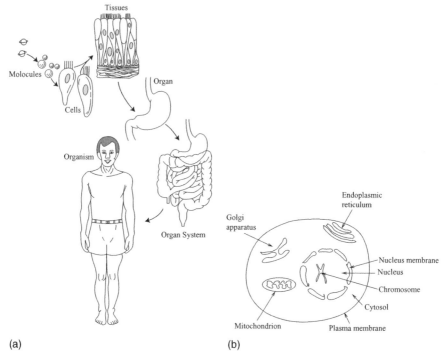

Figure 1.1   Schematic overview of an organism (a) and a eukaryotic cell (b). Figure (a) is reproduced from Norris and Siegfried (2011). This material is reproduced with permission of John Wiley & Sons, Inc.

Cell   The cell is the basic structural and functional unit of all organisms. From the *Gene Ontology (GO)*[1] definition it includes the plasma membrane and any external encapsulating structures such as the cell wall and the cell envelope, as found in plants and bacteria. There are hundreds of different types of mature cells, and in addition there are all the intermediate states that cells can take during development. Cells can have different shapes, sizes, and functions.

Tissue   A tissue is an interconnected collection of cells that together perform a specific function within an organism. All the cells of a tissue can be of the same type (a simple tissue), but many consist of 2–5 different cell types (a mixed tissue). Tissue types can be organized in a hierarchy, for example, the animal tissues are split into four main tissue types: epithelial tissues, connective tissues, muscle tissues, and nervous tissues. Blood is an example of a connective tissue. One assumes that there are less than 100 tissue (sub)types.

Organ   An organ is a group of at least two different tissue types such that they perform a specific function (or group of functions) in an organism. The

---

[1] www.geneontology.org.

boundaries for what constitutes an organ is debatable, but some claim that an organ is something that one can capture and extract from the body. Organs are also sometimes collected in organ systems.

Note that the above definitions are not without issues. For example it is not always clear whether 'something' is an organ or a mixed tissue. However, they should be sufficient for our use.

1.1.2 Composition of cells

The traditional way of looking at the organization of a (eukaryotic) cell includes two main concepts, *compartment* and *organelle*:

Compartment   A cell consists of a number of compartments (the term *cellular compartmentalization* is well established).

Organelle   One definition of organelle, Alberts et al. (2002) is: *membrane-enclosed compartment in a eukaryotic cell that has a distinct structure, macromolecular composition, and function. Examples are the nucleus, mitochondria, chloroplast, and the Golgi apparatus.* Thus an organelle is always a compartment.

In the most general understanding of these terms one can say that there is nothing in a cell that does not belong to a compartment, but there can be something that is not an organelle. An organelle consists of one or several compartments, for example, a mitochondrion has four.

The Gene Ontology project provides a more formal definition of the composition of a cell. The top level is the *cellular component*:

The cellular component ontology describes locations, at the levels of subcellular structures and macromolecular complexes. Examples of cellular components include nuclear inner membrane, with the synonym inner envelope, and the ubiquitin ligase complex, with several subtypes of these complexes represented.

Further it is noted:

Generally, a gene product is located in or is a subcomponent of a particular cellular component. The cellular component ontology includes multi-subunit enzymes and other protein complexes, but not individual proteins or nucleic acids. The cellular component also does not include multicellular anatomical terms.

Note that the cellular component not only includes the 'static' component of the cell, but also *multi-subunit enzymes and other protein complexes*.

The Gene Ontology now contains more than 2000 cellular components organized hierarchically as a directed acyclic graph.

It should be noted that compartment is not a central term in the Gene Ontology, but it appears in a couple of places as part of a component, for example, *ER-Golgi intermediate compartment* and *replication compartment*.

1.2   Homeostasis, physiology, and pathology

Homeostasis, *physiology*, and *pathology* are frequently used terms when describing protein quantification experiments. The following definitions are from Wikipedia:

Homeostasis   *is the property of either an open system or a closed system especially a living organism, that regulates its internal environment so as to maintain a stable, constant condition. Human homeostasis refers to the body's ability to regulate its internal physiology to maintain stability in response to fluctuations in the outside environment.*

Human physiology   *is the science of the mechanical, physical, and biochemical functions of humans in good health, their organs, and the cells of which they are composed. The principal level of focus of physiology is at the level of organs and systems. Most aspects of human physiology are closely homologous to corresponding aspects of animal physiology, and animal experimentation has provided much of the foundation of physiological knowledge. Anatomy and physiology are closely related fields of study: Anatomy, the study of form, and physiology, the study of function, are intrinsically tied and are studied in tandem as part of a medical curriculum.*

Pathology   *is the study and diagnosis of disease through examination of organs, tissues, bodily fluids, and whole bodies (autopsy). The term also encompasses the related scientific study of disease processes, called general pathology. Medical pathology is divided in two main branches, anatomical pathology and clinical pathology.*

1.3   Protein synthesis

The central dogma of molecular biology

$$DNA \rightarrow mRNA \rightarrow protein$$

gives a brief description of protein synthesis. DNA is located in the nucleus, and the mRNA is transferred to the cytoplasm, where protein synthesis takes place. The proteins can then again be transferred to other components, for example, to the nucleus or out of the cell.

1.4   Site, sample, state, and environment

Various molecular biology and medical experiments involve taking *samples* from a specific *site* from one or more research subjects. The subjects are most often of the same species, but cross-species experiments also occur.

A site is a specific 'part' of the subjects (one or several organs, tissues, or cell components), and is in one of several possible *states*. The state is specified by *features* (or *attributes*), and the value of each feature. How to describe the possible states, that is, which features to include, depends on the context. This means that which features to include depends on the goals of the analysis.

Example Suppose that we want to analyze how some properties of a site depend on time, for example, how the amount of proteins varies during the day. Then the (only) feature used is time. Another example is investigating the effects of using a given medicine over a certain time period. In this case the features are the amount of medicine and the time elapsed.

$\triangle$

When exploring how sites are affected by external stimuli, the features are often called *external features*.

The research subjects can be in a given *environment*, described by values of external features, for example, extremely low temperatures, resulting in the subject being in a frozen state.

1.5 Abundance and expression – protein and proteome profiles

Protein quantification concerns the determination of the amount of protein, relative, or absolute in gram or mole, in a sample of interest. In chemistry and biology however, *abundance* is used more or less synonymously with amount, and is in general the more popular term. We will therefore use abundance in favor of amount.

A related term is *concentration*, yet concentration is most often used to describe the ratio of the amount, mass, or volume of a component to the mass or volume of the mixture containing the component. For proteins, concentrations tend to be given in amount of protein per unit volume of mixture, for example, femtomole per microliter, $fm/\mu l$, given that most proteins are found in solution in living organisms. Note that for very low abundant proteins, the rather arbitrary unit of copy number per cell is sometimes used, for example, 100 copies per cell.

Another common term is *protein expression*. A strict understanding of this term in our context is *protein synthesis* (or *protein production*). It is however, also used for abundance, and the term *differentially expressed proteins* is often used for proteins with different abundances. We will here mainly use the term *differentially abundant*, to underline the difference between expression and abundance.

Protein abundances are commonly specified in *profiles*. We have two types of profiles, *protein profiles* and *proteome profiles*.

- A *protein profile* contains the abundances of a protein across a time series or across a set of sites/states, see Figure 1.2.

- A *proteome profile* consists of the abundances of a set of proteins from a site in a specific state.

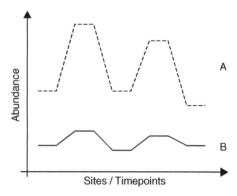

Figure 1.2   Illustration of the protein profiles for two proteins A and B.

The abundance of a protein can be determined for one research subject or for several subjects together (of the same species). If several subjects are included, the abundances are typically the average (or median) of the subjects, together with a measurement of the observed variation.

A protein profile thus contains the abundances of a single protein, while a proteome profile contains the abundances of a set of proteins.

Note that in the literature protein profile is sometimes (mis-)used for what we here define as a proteome profile. The understanding of the term should however be clear from the context in which it is used.

1.5.1   The protein dynamic range

The number of individual proteins occurring in a cell or a sample can vary enormously. It has been shown that in yeast the number of individual proteins per cell varies from fewer than 50 for some proteins to more than 10^6 for others. In serum, the difference between low abundance proteins and high abundance proteins can be from 10 to 12 *orders of magnitude*. One order of magnitude is ten to the first. The abundances of a specific protein can also vary enormously across sites/states.

To describe this variation the term *dynamic range* is often used. *Dynamic range* is a general term used to describe the ratio between the largest and smallest possible values of a changeable quantity. Dealing with large dynamic ranges is a challenge in protein quantification, and sets high requirements for the instruments used in order to be able to detect low-abundance proteins in complex samples.

1.6   The importance of exact specification of sites and states

Due to the large variations observed for protein abundances it is extremely important to clearly specify the sites where the samples were taken, and in which states the sites were, such that the resulting profiles reflect what one wants to investigate and compare. If not, one can end up comparing profiles where unconsidered features have influenced the profiles. Any detected differences may then come from these unconsidered features, and not from the features under consideration.

Many experiments try to analyze how the profiles depend on changing the value of one or more features of the sites. The challenge is then to keep all other features (that may influence the profiles) constant. However, simply finding these features in the first place can be difficult.

An additional challenge when comparing profiles from different sites, or from the same site in different states, is to try to extract the proteins under exactly the same experimental conditions, for example, using the same amount of chemicals and the same time slots. Neglecting this point may result in differences in the profiles (partly) due to different experimental conditions.

We here divide the features into five types, and give examples of each.

1.6.1 Biological features

These are features for which the values are more or less constant, or change in a regular manner, and can most often be easily measured. Examples are age, sex, and weight.

1.6.2 Physiological and pathological features

These are features that describe the physiological state of an individual organism, such as hungry, stressed, tired, or drained. Physiological features relate to healthy individuals, while pathological features relate to diseased individuals.

1.6.3 Input features

These are features characterizing the chemical and physical elements the individual has been exposed to, over longer or shorter time periods. Examples are food, drink, medicine, and tobacco. Make-up, shampoo, electromagnetic radiation, and pollution are also examples of input features.

1.6.4 External features

These are features describing the environment in which the individual lives, such as temperature, humidity, and the time of day.

1.6.5 Activity features

These are features describing the activities of the individual over longer or shorter time periods, such as different degrees of exercising, sleeping, and working.

1.6.6   The cell cycle

During its lifetime an individual cell goes through several phases. Four distinct main phases are specified for eukaryotic cells (G_1-, $S-$, G_2-, $M-$phase). The length of a cell cycle varies from species to species and between different cell types, from a couple of minutes to several years. Around 24 hours is typical for fast-dividing mammalian cells.

For some genes the expression level depends on which phase of the cell cycle the cell is in. This results in two types of proteins:

Dynamic proteins   are encoded by genes for which the expression level varies during the cell cycle.

Static proteins   are encoded by genes for which the expression level is constant during the cell cycle.

Due to the dynamic proteins, the proteome profile of a cell depends on the cell phase. It is often difficult, but not impossible, to detect this information at the individual cell level. During the last couple of years methods have been developed to make this possible. However, generally a large number of cells are extracted in a sample, and an average state of all these cells is commonly (implicitly) used, often assuming homeostasis.

1.7   Relative and absolute quantification

Generally there are two types of protein quantification, *relative quantification* and *absolute quantification*. Both relative and absolute quantification are used in proteomics, depending on the goals of the experiments.

Relative quantification   means to compare the abundance of a protein occurring in each of two or several samples, and determine the ratio of the occurrences between the samples.

Absolute quantification   means to determine the absolute abundance of a specific protein in a mixture. This can thus be used even when analyzing just a single sample.

Due to the properties of the instruments used it is generally easier to perform relative than absolute quantification.

1.7.1   Relative quantification

Relative quantification is primarily used for comparing samples to discover proteins with different abundances (differentially abundant proteins). Usually the set of proteins occurring in the samples are very similar, and the (large) majority of them also have similar abundances in all samples.

The basic approach is to compare two samples, and the aim is to find the relative abundance of each protein occurring in the samples, and mainly those proteins with different abundances.

Relative protein abundance can be specified in different ways. Let a_1 and a_2 be the calculated abundances of a protein in sample 1 and 2 respectively. Then the relative abundance can be presented as:

- **Ratio.** Defined as $\frac{a_1}{a_2}$.

- **Fold change.** Basically the same as ratio. However, another definition is also used: $\frac{a_1}{a_2}$ if $a_1 > a_2$ otherwise $-\frac{a_2}{a_1}$. This is symmetric, but results in a discontinuity at 1 and -1, which can be a problem for the data analysis.

- **Log-ratio.** $\log \frac{a_1}{a_2}$, also called *logarithmic fold change*.

Example Let the normalized abundances of three proteins in a protein sample S_1 be $(10, 20, 18)$ (of an undefined unit), and the same proteins in sample S_2 be $(40, 20, 17)$. Then there is a four-fold increase in S_2 for the first protein, and little or no change in the other two proteins.
△

1.7.2   Absolute quantification

Absolute quantification means to determine the number of individual molecules of specific proteins in a mixture. A common unit used is *mole*/volume (where one mole is $6.022 \cdot 10^{23}$, Avogadro's number). Also mass/volume is used. When the mass of a protein is known it is straightforward to convert between the two.

For proteins, one needs an instrument able to measure the number of copies of the specified protein in a mixture, which (for existing instruments) is more complicated than measuring relative abundances.

Note that absolute quantification could be used to calculate relative quantification. However, since absolute quantification is not straightforward and the results often have large uncertainties, this is not very common.

An example where the absolute abundance of a given peptide is of interest is when looking for *biomarkers*, see Chapter 18, where the absolute abundance of a peptide biomarker will provide useful information about the suitability of different assays to detect this peptide in a subsequent diagnostic procedure. Several (more or less similar) definitions of biomarkers exists. A biomarker in medicine is anything (substance or method) that can be used to indicate that a specific disease is present, or that an organism has a (high) risk of getting the specific disease. It may also be used to determine a specific treatment for a specific individual having a specific illness. A biomarker can include one or several proteins.

1.8   *In vivo* and *in vitro* experiments

In vivo means 'within a living organism,' and *in vitro* means 'within the glass.' Exactly what is meant by these terms in relation to experiments can vary. Studies using intact organisms, whether this be a mouse or a fruit fly, are always called *in vivo*. Studies where isolated proteins or subcellular constituents are analyzed in a tube, are always called *in vitro*. On the other hand, cell cultures grown in plastic dishes (a very common experimental technique) may be categorized as *in vivo* by some, while others, probably the majority, would define them as *in vitro*.

1.9   Goals for quantitative protein experiments

The overall goals for proteomics experiments are impossible to achieve with just one type of experiment. Several types of experiments (both small or large) can therefore be found. Examples of protein quantification goals are investigating how proteome or protein profiles:

- depend on changing states, for example, responding to different drugs, or different amounts of a drug;
- from similar states but varying sites differ;
- from corresponding sites from different categories of individuals differ, for example, between healthy and diseased persons;
- from corresponding sites from different species in similar states differ;
- from a site varies over time (under a constant environment);
- depend on the cell cycle.

A higher level goal is to explore if and how proteome or protein profiles can be used as biomarkers.

1.10   Exercises

1. Explain the difference between protein abundance and protein expression.

2. (a) Explain the difference between protein profile and proteome profile.

   (b) We have the following abundances (given in an arbitrary unit) for five proteins in four different cells:

	Cell 1	Cell 2	Cell 3	Cell 4
P_1	87	87	14	11
P_2	125	124	124	126
P_3	140	142	22	23
P_4	29	29	28	29
P_5	65	64	67	64

   Give examples of a protein profile and a proteome profile.

3. We have the following values for proteins in a cell: 21, 126, 2300, 560, 4700, 96 800, 19. What is the dynamic range? How many orders of magnitude are there?

4. Can you give examples of other features for each of the five types in Section 1.6?

5. A protein of mass 70 kDa has an absolute abundance of $5.1 \cdot 10^{10}$ pg/ml. What is the value in mole/ml? (Remember that 1 Da $= 1.66 \cdot 10^{-24}$ g, and that 1 mole $= 6.022 \cdot 10^{23}$.)

2

Correlations of mRNA and protein abundances

Proteins are the executive molecules of the cell, and it is therefore of vital interest to understand how the production and abundance of proteins depends on external stimuli, and how these changes are associated with diseases, cell differentiation, and other physiological and pathological processes.

Since mRNA provides the source template for protein production, it is worth investigating whether there is a relationship between the abundance of an mRNA-molecule and the abundance of the protein it encodes.

Note that we have chosen to use the term abundance for both mRNA and protein, although the term 'expression' is often used for the measured values of both mRNA and proteins. Combined expressions such as *comparisons of protein abundance and mRNA expression* are also often encountered.

2.1 Investigating the correlation

There are several reasons why investigating the correlation between mRNA abundance and protein abundance is interesting:

- If there is a correlation, many protein abundance experiments can be replaced by mRNA abundance experiments. Such mRNA experiments are easier to perform, and there is a lot of mRNA data already available in the public domain.

- Even if a general correlation cannot be found, correlations for certain sites in certain states may occur.

Computational and Statistical Methods for Protein Quantification by Mass Spectrometry,
First Edition. Ingvar Eidhammer, Harald Barsnes, Geir Egil Eide and Lennart Martens.
© 2013 John Wiley & Sons, Ltd. Published 2013 by John Wiley & Sons, Ltd.

- Even if no correlation can be found, mRNA experiments can give insight into the processes leading to protein synthesis.

In summary, we can claim that since proteins are the executive molecules in the cells, the main focus should be on the protein level. However, quantitative analyses on the mRNA level are easier to perform, and contribute to the exploration of the mechanisms of protein production. As a result, studies on the two levels are complementary rather than mutually exclusive, and analyzing the correlations of the different abundances can provide very interesting information.

A theory on how mRNA and protein abundances are related can be summarized as follows, Yu et al. (2007): *The protein synthesis rate is proportional to the corresponding mRNA concentration and the protein degradation rate proportional to the protein concentration.* This can be expressed in

$$\frac{dP}{dt} = k_s R - k_d P,$$

where

- P is the abundance of the protein under consideration;

- R is the corresponding mRNA abundance;

- k_s is the protein synthesis rate constant;

- k_d is the overall protein degradation and dilution rate constant.

In a steady state the change in abundance is zero, giving

$$P_\circ = \frac{k_s}{k_d} R_\circ = K R_\circ,$$

where $P_\circ$ and $R_\circ$ are the protein and mRNA abundances in the steady state, respectively.

Example Let $k_s = \frac{1}{20}$, $k_d = \frac{1}{40}$, and $R_\circ = 5000$. Then $P_\circ = 10\,000$. Note that during one time unit, $k_s R_\circ = 250$ of the individual mRNAs will code for a protein, and $k_d P_\circ = 250$ of the individual proteins will degrade, thus keeping the steady state.
$\triangle$

We can now investigate whether the constant $K = \frac{k_s}{k_d}$ is equal for some or all of the proteins, and whether it is constant for a given protein in different sites, or for the same site in different states.

To assure correct comparison, the mRNA and protein preparations must come from comparable samples. The easiest way of ensuring this is to use the same sample for both analyses. mRNA and proteins are extracted from the sample, and analyzed separately to determine their profiles. Typically, abundances are only used if they can be obtained for both the mRNA and the protein, thus avoiding missing data on either side.

Alternatively, techniques for handling missing data in the analyses must be used, see Chapter 8.3. The analysis usually relies on relative abundances, and the abundances have to be normalized, see Chapter 7.

2.2 Codon bias

Codon bias is a measure of the tendency of an organism to prefer certain nucleotide codons over others when more than one codon encodes for the same amino acid in a gene sequence. Indeed, codon bias exists because most of the amino acids can be encoded by several possible codons: there are 64 possible three-letter codons using the four bases A, C, T, and G, and these 64 codons are used to code for 20 amino acids and a stop signal for translation termination.

As a result, in the standard genetic code, L, R and S have six alternative codons each; A, G, P, T, and V have four alternative codons; I has three possible codons; C, D, E, F, H, K, N, Q, and Y all have two codons, while M and W only have one. There are also three stop codons that signal the end of protein synthesis. Since research has shown that codon bias affects mRNA and protein expression and abundance, we here provide a brief introduction to the topic.

Several measures for codon bias have been proposed, usually yielding a value between zero and one. As an illustration we here describe one of the initial methods, called *Codon Adaption Index (CAI)*, proposed in Sharp and Li (1987). CAI is calculated for a single gene, and is a measure of the degree in which usage of the most popular codons in the genome is reflected in that gene. Let G be the set of highly expressed genes in the considered genome, and let

- $n_{a,j}$ be the number of times that codon alternative j is used for amino acid a in G;

- y_a be the number of times the most used codon alternative for a is used in G;

- $w_{a,j} = \frac{n_{a,j}}{y_a}$ is then a weight showing the usage of coding alternative j relative to the most used alternative for the amino acid a.

CAI(g) is calculated over all codons in gene g as the geometric mean. Let

- m be the number of codons in g;

- w_i be the weight of the i'th codon in g.

Then

$$CAI(g) = \left(\prod_{i=1}^{m} w_i \right)^{\frac{1}{m}}.$$

Note that if the gene g uses only the most often used codon alternatives (in G), $CAI(g) = 1$. Any occurrence in g of rarely used codons in G will of course decrease the index.

CAI can be transformed (Jansen et al. (2003)) as follows. Consider all the 61 amino acid encoding genetic codons (thus excluding the three stop codons). Let k be an index representing these 61 codons, and let n_k be the number of times codon k appears in the gene g. Then

$$CAI(g) = \prod_{i=1}^{N}(w_i)^{\frac{1}{N}} = \prod_{k=1}^{61}\left(w_k^{n_k}\right)^{\frac{1}{N}} = \prod_{k=1}^{61} w_k^{\frac{n_k}{m}},$$

where $N = \sum_{i=1}^{61} n_i$.

Example Consider an example gene (where U is used for T) g=AUGUCCCCGAU CUUU, where the corresponding amino acids are MSPIF. Let us first calculate the weight for the second codon (UCC). Serine has four alternative codons, found to occur in the following numbers in the considered genes of the organism:

j=1	UCU:	1000
j=2	UCC:	744
j=3	UCA:	77
j=4	UCG:	17

Then $w_{S,2} = \frac{744}{1000} = 0.744 = w_2$. In the same way we can calculate the weights for the other amino acids (underlying data and calculations not shown): $w_1 = 1.0$, $w_3 = 1.0$, $w_4 = 1.0$, $w_5 = 0.296$. Then

$$CAI(g) = \left(\prod_{i=1}^{5} w_i\right)^{\frac{1}{5}} = 0.2202^{\frac{1}{5}} = 0.7389.$$

$\triangle$

Further variants of CAI are also proposed, see for example Carbone et al. (2003).

2.3 Main results from experiments

Numerous experiments have been performed to investigate the relation between mRNA and protein abundance, and a general correlation has been detected. It is however not strong enough to be used for individual proteins. It is therefore not possible to predict protein abundance from the corresponding mRNA abundance. This can partly be explained by the following:

- The RNA levels depend on the transcription efficiency and degradation rates of mRNA, while protein levels also depend on translational and post-translational mechanisms (including modification, processing, and degradation of proteins).

- The turnover rates (half-lives) for proteins (varying from a few seconds to several days) are generally longer than for mRNA.

- The uncertainty and possible experimental errors performed when determining the abundances, especially for proteins.

- The mRNA and the synthesized protein can (partly) be in different sub-cellular compartments. This must be taken into consideration if the comparison is to be performed for specific sub-cellular compartments.

However, several studies have shown significant correlations for more specific genes, sites, or states. Such correlations are important for our understanding of the dependencies and processes occurring in the cells.

Some studies have also shown that the correlation is higher when the codon bias is high (> 0.5).

2.4   The ideal case for mRNA-protein comparison

In the remainder of this chapter we will classify and briefly describe different types of correlation experiments for mRNA-protein abundances. The following *Entities* are used in the description: **gene, individual (or species), site, state, time point**.

Note that gene is here used for the DNA-origin of an mRNA and protein molecule (thus limited to the exons used in the actual transcript). Note also that we use *Entity* as a common term for all the items in the list, and that we use an upper case E to emphasize this specific use of the term.

In Tian et al. (2004) the ideal situation for an mRNA-protein comparison including genes and another Entity (for example, state) is described. The abundance data is represented in two tables $R_{i,g}$ for mRNA and $P_{i,g}$ for proteins, where $i = 1, \ldots, m$, with m the number of different instances of the Entity (for example, the number of different states), and $g = 1, \ldots, n$ with n the number of genes. Figure 2.1 illustrates this.

One has to make sure that corresponding columns in the two tables are abundances for the same gene, and that corresponding rows are from the same instance of the Entity, for example, in the same state. Corresponding rows in P and R should then

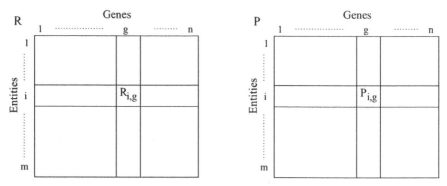

Figure 2.1   Tables for general comparison of mRNA and protein abundances.

be compared across the genes, and corresponding columns across the instances of the Entity. More advanced analyses could also be performed, such as pattern discovery or clustering.

Correlations are commonly illustrated using scatter plots, see Chapter 20.9. What the data points represent of course depends on the type of experiment.

2.5   Exploring correlation across genes

The question here is: *Is there a correlation between the mRNA and protein abundances across genes?* This corresponds to comparing two rows from the tables R and P respectively.

As an example see Figure 2.2 showing results from a comprehensive analysis of mRNA and protein abundances at steady state in Daoy medulloblastoma cells (tumor cells), Vogel et al. (2010).

It is important that the Entity values are described in detail, and that the experiments are designed to reveal biologically relevant information. Some important aspects are:

Individual: Are all the samples obtained from the same species?

Site: Are all the samples from the same organs, tissues or cell components?

State: Are all the samples from the sites in the same state?

Time point: Are all the samples from a single time point or spread across different time points?

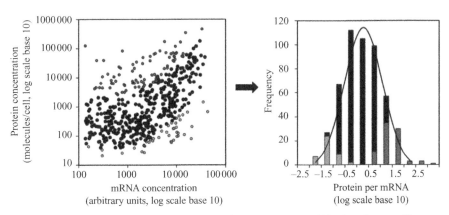

Figure 2.2   Left: A scatter plot of protein abundances to mRNA abundance of human cells at steady state, logarithmic scale. Right: Protein-per-mRNA (logarithmic scale) ratios, approximately log-normally distributed. Reprinted from Vogel et al. (2010) by permission of Nature Publishing Group.

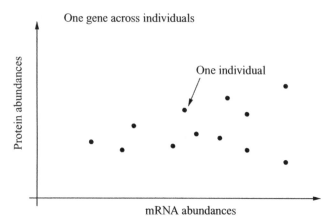

Figure 2.3   Comparison of mRNA and protein abundance for a single gene across individuals.

2.6   Exploring correlation within one gene

The question here is: *Is there a correlation between mRNA and protein for a single gene across a given Entity*? Here the Entity in question is varying, while all other Entities are fixed. This corresponds to comparing two columns in the tables R and P. Figure 2.3 illustrates such a comparison, with the individual as the varying Entity.

Remember that all Entities except the individual are fixed, but that these fixed Entities can be narrowly or widely defined, for example, the site can be a cell component (narrowly defined) or an entire organ (more broadly defined). The state can also be narrowly defined by specifying the value of all the features mentioned in Chapter 1.6, or broadly defined by using only one such parameter, for example, age.

Several studies have shown that the abundance ratio across individuals from the same species is fairly constant. If we know that this is the case for a specific gene, we can also use this to analyze variations that come from noise, or from variations and errors in the experiments.

Example If we know that the correlation of a specific gene is more or less constant, the data points in Figure 2.3 should form a straight line. Any deviations from this line can then be attributed to noise or to variations in the experiments.
△

Similar studies as illustrated in Figure 2.3 can be performed where the varying Entity is species, site, state, or time point.

2.7   Correlation across subsets

If a general correlation is not detected, one can try to explore if a correlation for specific subsets of the data can be found, for example, to see if there is correlation between subsets of the genes in Figure 2.2. However, given that a large number of genes are included in most such experiments, we can (almost) always find a set of genes that correlate in such a plot. The *p*-values for these subsets should therefore be calculated to assess their significance. These calculations are, however, not straightforward.

If the *p*-value is small, there is most likely a biological reason for the correlation, and one can try to examine which property is shared between the genes involved. An example of this is an experiment that found mRNA and protein abundance correlation only for proteins of high abundance.

2.8 Comparing mRNA and protein abundances across genes from two situations

We here consider mRNA abundances and protein abundances from two samples taken from two comparable conditions, for example, from two species or from two sites, or from two cell lines. This corresponds to comparing two rows from each of the abundance tables R and P, each row corresponds to one of the conditions.

Let the mRNA abundances for a given gene g be denoted $r_g^{(1)}$ and $r_g^{(2)}$ for the two conditions respectively, and $p_g^{(1)}$ and $p_g^{(2)}$ for the protein abundances. The question is then: *Does the change in mRNA abundance $r_g^{(1)} \rightarrow r_g^{(2)}$ reflect the change in protein abundance $p_g^{(1)} \rightarrow p_g^{(2)}$?*

In both quantitative mRNA expression analysis as well as in quantitative proteomics experiments, it is often the relative increase/decrease that is the most interesting. So instead of considering the differences $r_g^{(2)} - r_g^{(1)}$ and $p_g^{(2)} - p_g^{(1)}$ we consider

$$\frac{r_g^{(2)} - r_g^{(1)}}{r_g^{(1)}} \text{ and } \frac{p_g^{(2)} - p_g^{(1)}}{p_g^{(1)}}.$$

Since

$$\frac{r_g^{(2)} - r_g^{(1)}}{r_g^{(1)}} = \frac{r_g^{(2)}}{r_g^{(1)}} - 1$$

this is the same as considering

$$\frac{r_g^{(2)}}{r_g^{(1)}} \text{ and } \frac{p_g^{(2)}}{p_g^{(1)}}.$$

We can thus compare the changes (or differences) in mRNA and protein abundances for the two conditions using a scatter plot. We plot $\frac{r_g^{(2)}}{r_g^{(1)}}$ for each gene on the x-axis, and $\frac{p_g^{(2)}}{p_g^{(1)}}$ on the y-axis. However, as explained in Chapter 6.6, it is more

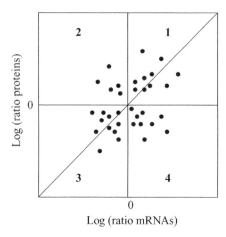

Figure 2.4   Scatter plot for comparing changes in mRNA and protein abundances. The numbers are the quadrant numbers.

convenient to take the logarithm of each fraction, thus defining the coordinates for each point as $(\log \frac{r_g^{(2)}}{r_g^{(1)}}, \log \frac{p_g^{(2)}}{p_g^{(1)}})$, as illustrated in Figure 2.4.

From such a scatter plot we can derive the following:

- a gene in quadrant 1 or 3 shows a positive relationship between mRNA and protein abundance changes;

- a gene in quadrant 2 or 4 has a negative relationship between mRNA and protein abundance changes;

- a gene on the line $y = x$ implies an equal relative change in both mRNA and protein abundance.

Correlation analysis can then be used to investigate if there is a relation between mRNA and protein abundance changes across the genes.

In Fu et al. (2007) an experiment is described that compares mRNA and protein abundances in humans and chimpanzees, indicating that there is a relation between mRNA abundance differences and the protein abundance differences. This conservation of abundance differences implies that characteristics of protein production from mRNA are retained between evolutionary related species.

2.9   Exercises

1. Consider a gene $g =$ AUGAUUGCAGGU. The following codon usage is observed in the genome:

	I		A		G	
j=1	AUU:	370	GCU:	3000	GGU:	1000
j=2	AUC:	2000	GCC:	366	GGC:	724
j=3	AUA:	6	GCA:	1758	GGA:	10
j=4			GCG:	1272	GGG:	19

Calculate the codon adaption index for the gene.

2.10   Bibliographic notes

Codon bias   Sharp and Li (1987), Carbone et al. (2003), Fox and Erill (2010).

Codon bias database   Hilterbrand et al. (2012).

Reviews   Maier et al. (2009)

mRNA protein correlation   Griffin et al. (2002), Greenbaum et al. (2003), Tian et al. (2004), Nie et al. (2005), Nie et al. (2006), Guo et al. (2008), Yu et al. (2007), Fu et al. (2007), Keshamouni et al. (2009), Pascal et al. (2008), Vogel et al. (2010), Schwanhäusser et al. (2011), Maier et al. (2011).

Programs for mRNA protein comparison   Yu et al. (2007) (http://proteomics .gersteinlab.org).

3

Protein level quantification

Quantitative proteomics experiments have been performed for many years, using different methods and techniques. In this chapter we briefly describe the most common methods that do not rely on the combination of liquid chromatography and mass spectrometry.

3.1 Two-dimensional gels

In general, 2D SDS-PAGE gel quantification is based on the signal intensity of the spot in which the protein has been found. It is important to note that this intensity is an indirect measure, since protein spots are first stained or labeled in order to become visible, and it is the intensity of the stain or label that is subsequently measured. Obviously, spots containing more than one protein present challenges to any such quantitative study as the unexpected proteins add staining intensity that will be incorrectly assumed to be derived from the identified protein. And even if the spot is known to hold more than one protein, it is not possible to deconvolute the contributions of the individual proteins to the total amount of staining.

On the other hand, different variants such as truncated or modified forms of the protein are typically spread over different spots, making the global quantification of a protein's expression level problematic. Indeed, such an analysis requires the addition of all intensities across all the different variant spots. Missed spots, that is, spots that could not be identified for whatever reason, will thus result in an underestimation of the quantification. Also remember that each individual spot has a chance of containing contaminant proteins that can contribute to staining intensity. The fact that different forms often localize in different spots can however be beneficial if one wants to compare different variants with one another, for example, comparing a phosphorylated form of a protein to its unphosphorylated form.

Computational and Statistical Methods for Protein Quantification by Mass Spectrometry,
First Edition. Ingvar Eidhammer, Harald Barsnes, Geir Egil Eide and Lennart Martens.
© 2013 John Wiley & Sons, Ltd. Published 2013 by John Wiley & Sons, Ltd.

Another difficulty for 2D gel based quantification lies in the usable range of the staining procedures used for quantification. Several staining procedures do not stain at all below a certain amount of protein present, and become saturated above a certain amount of protein present. The interval between these amounts can be considered 'covered' by the staining procedure and is referred to as the *dynamic range*.

In addition to high sensitivity, it is desirable to have a linear response in staining intensity versus amount of protein present for the dynamic range. Most nonfluorescent staining agents have relatively poor dynamic range, with even the best nonfluorescent technique (silver staining) only yielding a dynamic range of about one order of magnitude. Fluorescent stains can attain the sensitivity of silver staining when used under optimal conditions, and can result in a much better dynamic range, providing up to five orders of magnitude. The fluorescent detection does come with its own caveats however, including fading signals if the fluorophore is (slowly) decomposed by the influx of light, and problems with the interference of fluorescence of flourophores that become associated with detergent micelles rather than proteins.

3.1.1   Comparing results from different experiments – DIGE

One of the most serious issues of 2D gel based quantification lies in the poor re-producibility of gels. As a result, comparing gels is not a simple process. A widely-used technique to avoid the typical low reproducibility between different gels is the fluorescence-based Differential In-Gel Electrophoresis (DIGE) approach. In this technique, the proteins in the different samples are labeled with different fluorescent dyes having different excitation wavelengths. The samples are then mixed, and the proteins separated on the same 2D PAGE gel. Because of the different excitation wavelengths of the labels, separate gel patterns can be obtained for each sample. This effectively allows the samples to be run under identical circumstances, greatly limiting the factors that can cause variation, and the spots from two samples are therefore much more directly comparable. Figure 3.1 illustrates the process.

3.2   Protein arrays

Another often-used method for discovery oriented protein quantification is provided by protein arrays. The use of protein arrays can be explained by considering three components:

- a glass plate or slide in a regular grid;
- a set of capture molecules that are fixed to the plate;
- a set of binding molecules that are to be bound to the capture molecules.

The capture molecules are typically antibodies, proteins, or peptides, although other (macro-)molecules such as DNA can be used as well. The different types of capture molecules enable different types of analyses to be carried out. We can divide the protein arrays into forward arrays and reverse arrays.

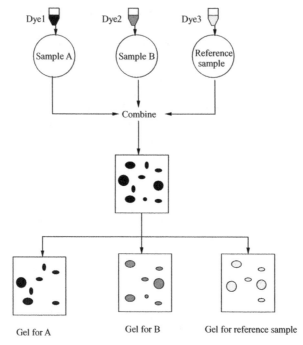

Figure 3.1   Illustration of the DIGE procedure. Two samples plus a reference sample (usually obtained by pooling) are each labeled with different dyes and then mixed. A 2D gel is then used to separate the proteins. Using different excitation wavelengths of light, three different gel images can be obtained from the same gel, one for each sample.

3.2.1   Forward arrays

In forward arrays the proteins to be analyzed are the binding molecules.

Antibody arrays

In this design the capture molecules are a collection of antibodies. Upon sample application, these antibodies bind the proteins they were raised against, and bound proteins can subsequently be detected. Despite the straightforward design, there are several problems with this type of array. First and foremost, there must be a sufficiently specific and sensitive antibody available for each protein to be measured. This is for instance not the case for every human protein, let alone for less well-studied species. Furthermore, antibodies tend to be expensive, and can often suffer from a specific binding (leading to a falsely exaggerated signal) or poor binding efficiency (leading to a falsely subdued signal). Indeed, antibodies that function quite well in simpler situations such as Western blotting (next Section), may not function as well when confronted with a whole-proteome mixture. Since proteins cover a very broad range of physiochemical parameters, it is furthermore difficult to optimize binding and washing conditions to ensure reasonable capture of all proteins.

Other types of forward arrays

Since antibody arrays can be challenging from a practical perspective, other types of forward arrays have been developed, where the capture molecules consist of peptides, small molecules or stretches of DNA sequence. Usually, these arrays are designed to capture a specific subset of proteins, such as DNA-binding, or drug-binding proteins.

3.2.2   Reverse arrays

As the name implies, reverse arrays take the opposite approach, and the proteins to be analyzed are the capture molecules. The objective here is often antibody profiling, where a sample of plasma or serum from a patient is run over the array to detect antibodies against the spotted proteins. Such arrays are particularly useful for the detection of auto-antibodies, that are aimed at the patient's own proteins and form important factors in a variety of auto-immune diseases. These arrays are more easily produced than antibody arrays, but still pose the challenge that the relevant proteins must all be (recombinantly) expressed and purified. Reverse arrays that cover most of the proteome of various model organisms, and of course humans, are now commercially available.

3.2.3   Detection of binding molecules

Detection of the binding molecules is usually carried out using fluorescence, although more advanced techniques such as surface plasmon resonance and atomic force microscopy can also be used. In the case of fluorescent detection, the fluorophore can either be attached directly to the sample proteins after lysis, or it can be added after capture through the use of a secondary antibody in so-called sandwich designs.

3.2.4   Analysis of protein array readouts

Since protein arrays closely resemble RNA microarrays, the processing of readout data is commonly handled in the same fashion. Protein arrays typically follow design guidelines copied from RNA microarrays as well, including the use of on-array replicate spots that are scattered over the array.

3.3   Western blotting

Where 2D gel electrophoresis and protein arrays provide discovery oriented methods used to detect and quantify as many proteins as possible in a sample, Western blotting provides a targeted means to quantify a single protein in a sample. In Western blotting, the protein complement of a sample after lysis is typically separated on a single gel dimension, and the separated proteins are subsequently transferred (blotted) onto a membrane with a certain affinity for proteins (typically nitrocellulose or polyvinylidene fluoride (PVDF)).

This membrane is then probed with an affinity reagent that binds specifically to the protein of interest. This affinity reagent is most commonly an antibody specific to the protein. Detection of the bound antibody can be performed through a secondary antibody carrying a linked enzyme, or directly through a linked enzyme on the primary antibody.

The linked enzyme is chosen so it performs an observable reaction, most commonly horseradish peroxidase cleaving a chemoluminescent reagent, producing detectable light in proportion to the amount of bound enzyme, in turn dependent on the amount of protein. Other types of detection are possible, for instance using antibodies connected to a fluorophore, and the use of radioactively labeled antibodies. This last method however, is expensive, laborious, and dangerous, and is only used in those cases where exquisite sensitivity is required.

3.4   ELISA – Enzyme-Linked Immunosorbent Assay

Another popular means of targeted protein quantification is offered by ELISA (Enzyme-Linked Immunosorbent Assay). Here, a whole proteome lysate is affixed to a solid substrate (often a microtiter plate) and is then probed by a specific affinity reagent, usually an antibody. This affinity reagent is then either directly or indirectly (via a secondary antibody for instance) detected through a coupled enzyme (often horseradish peroxidase, similar to Western blotting).

ELISA is often used in clinical assays, and the development of an ELISA is therefore often the endpoint of a biomarker discovery pipeline.

3.5   Bibliographic notes

2D gel analysis   Rabilloud et al. (2010).

Protein arrays   Cahill and Nordhoff (2003), Maercker (2005), Wilson et al. (2010), Kijanka and Murphy (2009), Reid et al. (2007).

ELISA   Zangar et al. (2006), Zangar et al. (2004).

4

Mass spectrometry and protein identification

The use of mass spectrometry is now the dominant approach for protein quantification. In most protein quantification experiments the identity of the proteins is unknown, and protein identification is necessary. This chapter provides a brief overview of the principles of using mass spectrometry for protein identification and characterization. More extensive introductions can be found elsewhere, for example in Eidhammer et al. (2007).

Various protein properties are used in protein identification, perhaps most important is the mass. The protein mass is constituted by the residue masses. Table 4.1 shows the main properties of the 20 standard amino acids.

4.1 Mass spectrometry

Generally there are two paradigms for proteomics by use of mass spectrometry (MS), *top-down* and *bottom-up*. Top-down uses mass spectrometry on the intact proteins, while bottom-up cleaves (digests) the proteins into *peptides* before performing mass spectrometry.

Cleavage is performed using proteases, most often trypsin which cleaves after each arginine and lysine unless followed by a proline. If the first amino acids in a protein are MATVLLPQRILVSTSTEKWETPWYKA..., the first peptides will be MATVLLPQR, ILVSTSTEK, WETPWYK. Using trypsin normally results in peptides of length 6–20 amino acids, which is suitable for mass spectrometry.

Computational and Statistical Methods for Protein Quantification by Mass Spectrometry,
First Edition. Ingvar Eidhammer, Harald Barsnes, Geir Egil Eide and Lennart Martens.
© 2013 John Wiley & Sons, Ltd. Published 2013 by John Wiley & Sons, Ltd.

Table 4.1 The main amino acid properties. The residue masses are from the i-mass guides (http://i-mass.com/guide/aamass.html), pI values from Nelson and Cox (2004), pK values from DTASelect (http://fields.scripps.edu/DTASelect), hydrophobicity indexes from Kyte and Doolittle (1982), and average occurrences from the composition of the Swiss-Prot database release 2012_05.

			Residue mass		pI values	Hydro- phobicity index	pK values	Occurrence (%)
Amino acids			Monoisotopic	Average				
Ala	A	alanine	71.04	71.08	6.01	1.8		8.26
Arg	R	arginine	156.10	156.19	10.76	−4.5	12.0	5.53
Asn	N	asparagine	114.04	114.10	5.41	−3.5		4.06
Asp	D	aspartic acid	115.03	115.09	2.77	−3.5	4.4	5.45
Cys	C	cysteine	103.01	103.15	5.07	2.5	8.5	1.36
Gln	Q	glutamine	128.06	128.13	5.65	−3.5		3.93
Glu	E	glutamic acid	129.04	129.12	3.22	−3.5	4.4	6.75
Gly	G	glycine	57.02	57.05	5.97	−0.4		7.08
His	H	histidine	137.06	137.14	7.59	−3.2	6.5	2.27
Ile	I	isoleucine	113.08	113.16	6.02	4.5		5.96
Leu	L	leucine	113.08	113.16	5.98	3.8		9.66
Lys	K	lysine	128.10	128.17	9.74	−3.9	10.0	5.84
Met	M	methionine	131.04	131.20	5.74	1.9		2.42
Phe	F	phenylalanine	147.07	147.18	5.48	2.8		3.86
Pro	P	proline	97.05	97.12	6.48	−1.6		4.70
Ser	S	serine	87.03	87.08	5.68	−0.8		6.55
Thr	T	threonine	101.05	101.11	5.87	−0.7		5.34
Trp	W	tryptophan	186.08	186.21	5.89	−0.9		1.08
Tyr	Y	tyrosine	163.06	163.18	5.66	−1.3	10.0	2.92
Val	V	valine	99.07	99.13	5.97	4.2		6.87
N-terminus			1.01*				8.0	
C-terminus			17.01*				3.1	

*Note that the masses for the termini are the masses of the atoms added to the end residues, normally *H* at the N-terminus and *OH* at the C-terminus.

The currently dominating paradigm is the bottom-up approach, and this is therefore the focus of this introduction. Bottom-up can further be divided into two approaches, *Peptide Mass Fingerprinting – PMF* and *MS/MS* (or *tandem MS*). They differ in the peptide properties used, the mass spectrometry instruments applied, and the methods used to separate molecules (peptides or proteins) to reduce complexity.

4.1.1 Peptide mass fingerprinting (PMF)

Peptide mass fingerprinting first separates the intact proteins, mainly using 2D gel electrophoresis, which separates the proteins according to mass and isoelectric point (pI).

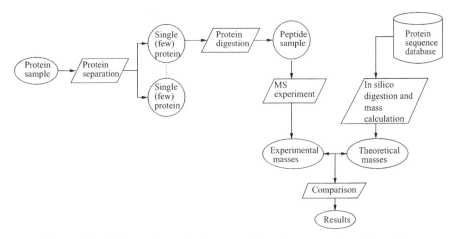

Figure 4.1   MS protein analysis approach, using separation of proteins.

The ideal separation should result in all molecules of a single protein ending up in a single fraction. In practice however, protein separation techniques rarely achieve this ideal resolution, especially if the starting material is a complex mixture containing many proteins. For example the variants of a single protein, such as modified proteins, will usually end up in different fractions. However, if each fraction contains only a small number of proteins (two or three), the separated fractions may still be manageable for further analysis by mass spectrometry.

The approach is illustrated in Figure 4.1.

4.1.2   MS/MS – tandem MS

The MS/MS approach first digests the protein sample into peptides, and is also referred to as *peptide-centric proteomics*. A relatively simple protein sample such as an *E. coli* cell lysate, will contain somewhere between 2500 and 5000 proteins. With an average of 25 peptides per protein in *E. coli*, this results in 62 500 peptides if 2500 proteins are expressed.

Thus a separation of the peptides is needed before the mass spectrometry analysis. This is performed by chromatography, mainly using liquid chromatography (LC) or high performance LC (HPLC), where the separation is mainly based on hydrophobicity, charge, or size. The approach is illustrated in Figure 4.2. For large scale proteomics two dimensions of separation is often used, for example, MudPIT where a separation on charge is followed by a separation on hydrophobicity.

4.1.3   Mass spectrometers

Mass spectrometry is used to measure the mass, or more correctly the mass-over-charge ratio (m/z), of the components in a sample. The instruments are called *mass spectrometers*.

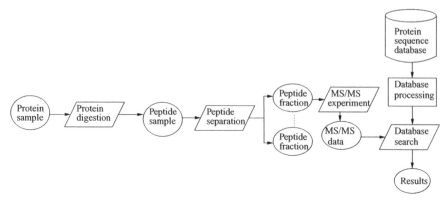

Figure 4.2   MS/MS protein analysis approach, using separation of peptides.

A mass spectrometer consists of three main parts: The *ionization source*, the *mass analyzer*, and the *detector*, see Figure 4.4. All mass spectrometers use electric or electromagnetic fields to handle the components of the sample. The components being analyzed therefore have to be ionized before their masses can be measured, and this is done in the ionization source. Ionization is commonly achieved by the addition of protons to the molecules, which also increases the mass of the molecules by the nominal mass of 1 Da per charge (per proton).

The sample is then transferred to the mass analyzer which separates the components of the sample according to the mass-over-charge ratio (m/z) of the ions. After separation the components hit the detector, and a *mass spectrum* is constructed by a connected computer. A mass spectrum is a diagram, with the m/z values along the horizontal axis, and the intensity values of the signal for each component along the vertical axis.

Since the analyzers operate on m/z rather than on the mass directly, the charge of a component must be known before the mass can be determined. m/z is usually reported as a dimensionless number, but units like *Thomson* (Th), u, or even Da are sometimes used. Note that the use of u or Da is essentially incorrect, whereas Thompson is correct, but not recognized by standardization bodies.

Example Figure 4.3 illustrates the main principles of mass spectrometry.

Suppose that we have a sample of three peptides (a, b, c), with nominal masses a: 680, b: 481, c: 400. Assume further that some of the peptides become singly charged, and some doubly charged, while all b peptides are singly charged, and all c peptides doubly charged. The m/z values for the ions thus become a:$(680 + 1)/1 = 681$ and $(680 + 2)/2 = 341$, b:482, c:201, as shown in the lower right spectrum in Figure 4.3.

$\triangle$

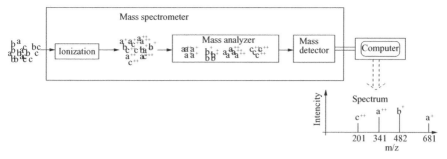

Figure 4.3   The main principles of mass spectrometry. In the ionization source, the sample components are transferred to gas phase and acquire their charges. In the mass analyzer the components are separated according to their m/z values before hitting the detector. A computer connected to the instrument constructs the mass spectrum. See the example in the text for more details.

Ionization sources

As mentioned at the beginning of this chapter, there are two main classes of mass spectrometers for proteomics: those that perform single mass spectrometry and those that perform tandem mass spectrometry (MS/MS).

The former measures the m/z ratio of intact peptides, and for this it is an advantage if the peptides mainly have a single (positive) charge. In tandem mass spectrometry the peptides are intentionally fragmented (mainly due to peptide backbone breaking, resulting in two fragments per fragmentation event) to also measure the masses of fragments. Ideally, one wants to detect the mass of both fragments, which can only be achieved if both are ionized. Thus, in tandem MS it is an advantage if the peptides carry several charges (often $+2$) as this increases the chance that all of the fragments will be ionized. The two classes of instruments therefore generally use different ionization sources.

MALDI and ESI are the most common ionization sources for proteomics. MALDI almost exclusively results in a single charge, while ESI most often produces two to four charges.

Mass analyzers

There are several techniques employed for mass analyzers used in proteomics, here we briefly describe the simplest alternative.

In the MALDI *time-of-flight (TOF)* mass analyzer, ions are accelerated by an electric field in the ionization source, and then enter a field-free drift tube. The velocities the ions have achieved during the acceleration depend on the mass and the charge of the ions, and these velocities are kept during the travel through the drift tube.

Naturally, the time needed to pass through the drift tube depends on the velocity. When the ions hit the detector at the end of the drift tube, the time of flight is

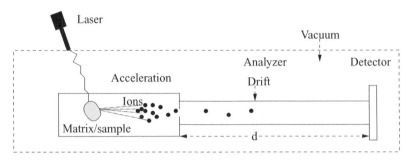

Figure 4.4   Illustration of a linear MALDI-TOF.

registered, and the m/z values can be calculated. The principle for a linear TOF analyzer is illustrated in Figure 4.4.

4.2   Isotope composition of peptides

Many elements naturally exist in several isotopes. When these elements are incorporated into biological molecules, such as amino acids, the isotopes are represented in ratios that correspond to their abundance in nature. Since the chemical properties of the different isotopes of an element are identical, the chemistry of the biomolecules (and therefore their function) is not affected by isotope differences.

Six elements (hydrogen, carbon, nitrogen, oxygen, phosphorus, and sulfur) constitute the overwhelming majority of the elements incorporated into proteins and their post-translational modifications. The abundances and masses of the stable isotopes of these six elements are shown in Table 4.2. Note that several elements also have radioactive isotopes, for example, ^3H and ^{32}P, which are excluded from this discussion as they are subject to decay over time.

Numerically, carbon and hydrogen are the most common elements in proteins, but as the heavy isotope of carbon, ^{13}C, has a 100-fold higher abundance than deuterium, ^2H, the former has a far greater influence on the isotope pattern of peptides than any of the other elements. We will therefore focus on isotopes of carbon, and assume that all the other elements are always represented by their lightest isotope.

Assume that we are working with a small peptide with a mass of approximately 600 Da (referred to as mass M). Such a peptide contains around 30 carbon atoms. Approximately one third of the peptide molecules will therefore contain one ^{13}C atom with the remaining being ^{12}C. Given that only a very small amount of the peptides contains two (or more) ^{13}C atoms, approximately two thirds of the peptide molecules will only contain ^{12}C.

As there is a 1.0034 Da mass difference between the peptide molecules that only contain ^{12}C and the peptide molecules that contain one ^{13}C atom, the mass spectrum will include one high peak at mass M, which is called the *monoisotope peak*. At mass M + 1 there will be a peak of approximately one half the intensity of the monoisotope

Table 4.2   Abundances and masses of the stable isotopes of elements naturally occurring in proteins and their post-translational modifications.

Elements		Abundances (%)	Masses
Hydrogen	^1H	99.99	1.0078
	^2H	0.01	2.0141
Carbon	^{12}C	98.91	12.0000
	^{13}C	1.09	13.0034
Nitrogen	^{14}N	99.63	14.0031
	^{15}N	0.37	15.0001
Oxygen	^{16}O	99.76	15.9949
	^{17}O	0.04	16.9991
	^{18}O	0.20	17.9992
Phosphorus	^{31}P	100.00	30.9738
Sulphur	^{32}S	95.02	31.9721
	^{33}S	0.76	32.9715
	^{34}S	4.22	33.9676

peak. A very small peak may also be seen at mass $M + 2$, corresponding to the peptide with two ^{13}C atoms, as Figure 4.5 illustrates.

If we instead are working with a peptide of mass approximately 3000 Da, there will be a low percentage of the peptide molecules that only contain ^{12}C atoms. A higher percentage of peptide molecules will contain one or two ^{13}C atoms. Peptide molecules containing three, four, or five ^{13}C atoms can be seen in decreasing amounts.

Figure 4.5 illustrates the different behavior of the isotope peaks of a relatively small and a relatively large peptide. Thus, with increasing mass of the peptide, fewer

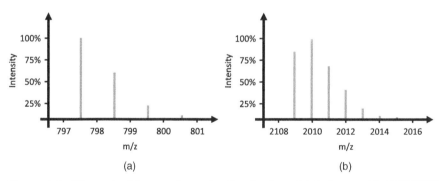

Figure 4.5   Theoretical isotope intensity distribution of (a) the peptide FIAVGYK (monoisotope mass 796.448, average mass 796.953), and (b) the peptide TQT-WAGSHSMYFYTSVSR (monoisotope mass 2107.942, average mass 2109.277). The masses are without charge.

and fewer molecules will contain only ^{12}C atoms. For peptide masses above 5 to 7000 Da, most instruments will no longer be able to distinguish the monoisotope peak, and the other peaks will be poorly resolved. In experimental settings, we therefore rarely use the monoisotope mass of large peptides or proteins (although it can easily be calculated), but rather the average mass, calculated as a weighted average over the isotope masses.

A collection of isotope peaks from the same peptide is called an *isotope pattern* or an *isotope envelope*. Knowledge of the isotope peak pattern helps to interpret mass spectra and is essential for many of the methods used for protein quantification.

It is often useful to represent an isotope pattern by a single peak, with intensity equal to the sum of the isotope intensities. *Monoisotoping* reduces the pattern to the peak with lowest m/z of the individual isotopes. *Deisotoping* reduces the isotope pattern to a centroid peak, with the m/z value determined from the intensities of the individual isotopes.

4.2.1   Predicting the isotope intensity distribution

There are a lot of tools for calculating the intensity distribution over the isotope pattern for a given chemical formula. When the sequence of a peptide is known, it is therefore possible to calculate the theoretical intensity distribution, as shown in Figure 4.5.

A common task in proteomics using mass spectrometry is to try to reveal the isotope patterns, and a theoretical intensity distribution can be of great help. The chemical formula of the peptide under consideration is however not known since the sequence is unknown, only the m/z. However, using the expected number of residues in a peptide of given mass, and the expected number of carbons in a residue, the isotope intensity distribution can be estimated. A set of peaks in a spectrum can therefore be compared to the expected distribution of a peptide of that mass to investigate if the set can constitute a specific isotope pattern.

4.2.2   Estimating the charge

Singly charged peptides will have a delta, that is, a difference, of $(m/z) = 1$ between the peaks in the isotope pattern as seen in Figure 4.5(a). Doubly charged peptides will have a delta of $(m/z) = 0.5$ between the peaks, and triply charged peptides will have a delta of $(m/z) = 0.33$.

At the same time, the m/z position of the peak in the mass spectrum is moved according to the formula $(M + zH^+)/zH^+$, where M is the mass of the peptide. This can be used to estimate the charge of peptides.

4.2.3   Revealing isotope patterns

Revealing isotope patterns in a spectrum is complicated by overlapping patterns, as illustrated in Figure 4.6.

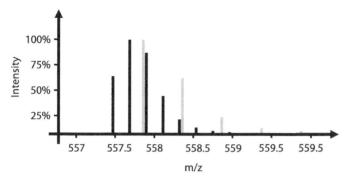

Figure 4.6   Example of overlapping isotope patterns from two different peptides, LHHVSSLAWLDEHTLVTTSHDASVK (charge 5) and LGINSLQELK (charge 2). The heaviest peptide has eight isotopes (black), and the lightest peptide has five isotopes (light grey).

Programs for splitting a set of peaks into different patterns should at least use the two following constraints to identify isotope patterns:

- The difference d between two neighboring isotope peaks in a pattern must satisfy:

$$\left| d - \frac{\Delta m}{z} \right| \leq \delta,$$

  where

  - Δm is the expected mass difference between two isotopes of a peptide. This depends on the mass of the peptide, but a representative value is used, for example, $\Delta m = 1.00286864$ (Cox and Mann (2008)), the mass difference between the ^{13}C peak and the monoisotope peak in an averaging peptide of 1500 Da mass.

  - δ is the accuracy of the comparison. It can be a fixed value, or a function of the standard errors of the two measured peak masses, and a component related to the mass shift a sulfur atom can cause (Cox and Mann (2008)).

  - z is the charge, that will be in a range from one to a maximum value for the number of charges that a peptide can have in the actual experiment.

- The intensities of the peaks should correlate with the expected intensity distribution of peptides of the actual mass. The Pearson correlation coefficient can for example be used to calculate the correlation, see Chapter 20.8.

A program for splitting and revealing the different isotope patterns can be based on using graphs with the peaks as nodes, and an edge between peaks that can be neighbors in an isotope pattern. Consistent connected subgraphs can then be found, where consistent means that the edges must be based on the same charge, and the patterns can correlate with the theoretical distributions.

For MS instruments of low or medium mass accuracy one peak can represent more than one isotope, and this must then be taken into consideration in the program.

4.3   Presenting the intensities – the spectra

A basis intensity detection is done by the detector, and the result is called the *detector trace*. The detector trace is however commonly treated before being presented to the user, either as *raw data* or as a *peak list*.

Raw data   Raw data is the output file from the instrument, and is the most similar to the detector trace. A detected molecule is presented in the spectrum as a peak stretching over a (short) range of m/z values, rather than a single bar at an exact m/z value.

Thus for each peak there are numerous intensity measurements at defined small increments of the m/z value, defined by the resolution time of the detector. These files are large, typically several gigabytes. Figure 4.7 shows an example of a raw data spectrum with little noise.

A spectrum in a raw data file is said to be in *profile mode*.

Peak list   The huge number of measurements in the raw data makes it inappropriate for most succeeding treatments, and the spectra are therefore usually transformed to peak lists, where all the measurements for each peak are collected into a single mass value. This value is usually the intensity value at the centroid of the peak.

In its simplest form the peak list is only a list of m/z values of the detected peptides. Most peak list formats, however, also include the intensity value and possibly additional data, for example, the charge state for each detected peptide. The intensity (the number of detected ions) is proportional to the area under the raw data curve. A spectrum in a peak list file is said to be in *centroid mode*.

Processed peak list   A peptide can occur as several isotopes and with several charges. It is often desirable to collect these into a single peak, thus *deisotoping and charge deconvolution* (to charge one) is performed.

A typical raw data spectrum can contain about 3000 value pairs (m/z, intensity) inside a 100 unit interval, while the derived peak list may contain only four pairs (corresponding to four peptides). The ratio between these numbers depends on the instrument used.

Transforming raw data to peak lists is a nontrivial transformation, and is partly discussed in the next section.

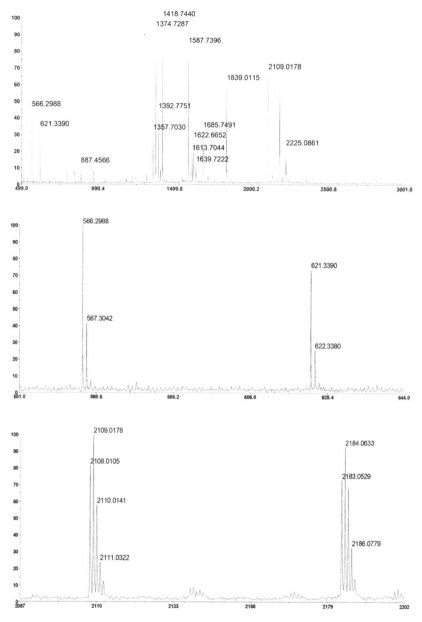

Figure 4.7   Example of a peptide mass fingerprint of a tryptic digest of the matrix metalloproteinase 2 protein. For ease of viewing, not all peaks have been labeled in the top spectrum. Two peaks from the lower and two peaks from the higher mass range of the spectrum are shown in the two zoomed-in views below. Note the differences in the isotope intensity distribution between the lower mass and the higher mass peptides. Also note that the isotope peaks have a spacing of approximately 1, indicating that all of these peptides carry a single charge.

4.4   Peak intensity calculation

Quantification of proteins is based on the intensities of the peaks in the spectra, thus high quality intensity calculation is required, and this calculation should be based on the raw data.

Before area calculation the peaks have to be detected, which can be done by standard peak detection programs. Additionally the *baseline* should be subtracted, the baseline being the measured intensities when no compound is present.

The calculated intensity should be as correct and precise as possible. This is especially important since errors made at this stage will propagate in the later analyses. In some programs the errors (uncertainties) are estimated and used in the succeeding calculations.

Different ways of performing intensity determination exist, and many of the programs for mass spectrometry offer several options. We will here mention three types of methods, where the last two use the area under the curve fitted to the data.

Use of the apex   The apex intensity, that is, the top of the peak, is used.

Quadratic fitting   Three points are chosen, the apex and one point on either side of the apex. The unique parabola through the points is then calculated (quadratic curve fit).

Gaussian or Lorenzian curve fitting   The raw data measurements are fitted to a Gaussian or a Lorentzian curve, going through the apex. Generally a Lorenzian curve is wider then the corresponding Gaussian curve (when they have for example equal half-line width at half-maximum intensity).

Gaussian and Lorenzian curves are symmetrical, while the quadratic fitting does not have to be.

4.5   Peptide identification by MS/MS spectra

In most quantification experiments the proteins in the samples are unknown, and protein identification is usually included as part of the experiment. We therefore give a brief introduction to the task of protein identification, a more thorough description can be found in Eidhammer et al. (2007).

Proteins are identified based on peptide identifications. However, going from identified peptides to proteins is not straightforward, and is referred to as the *protein inference problem*. We often say that determining which proteins are in a sample is done by *inferring* the proteins from the identified peptides. The protein inference problem is briefly treated in the Section 4.6.

An MS/MS spectrum for a peptide is created by splitting the peptide into smaller fragments, and producing a spectrum of the fragments being ionized. Each fragment may obtain one or several charges, depending on the charge of the peptide.

The peptides are mainly fragmented along the peptide backbone. If a charge is retained on the N-terminal fragment, the fragment ion is classified as either *a*, *b*,

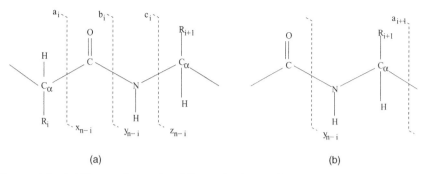

Figure 4.8   (a) Nomenclature for different backbone fragments. R denotes residue, n is the number of residues in the peptide, i refers to the i-th residue of the peptide. (b) The basis of an immonium ion.

or c, depending on which bond is broken. If a charge is retained on the C-terminal fragment, the fragment type is either x, y, or z. Fragment ions are annotated by subscripts, indicating the number of residues in them, see Figure 4.8(a). Additional fragment ion types can also occur, for example, immonium ions, see Figure 4.8(b).

An example of an experimental MS/MS spectrum is shown in Figure 4.9. The peptide origin of an MS/MS spectrum is termed the *precursor*.

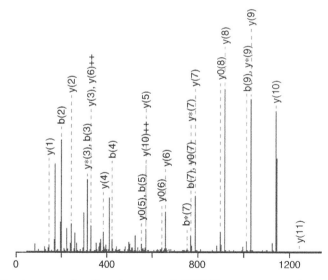

Figure 4.9   An example of an interpreted MS/MS spectrum as shown in Mascot output. Many fragment ions are observed, for example all singly charged y-ions are recognized. Water loss (marked by 0) and ammonia loss (marked by *) are recognized for some fragments. Note that in practice b_1-ions are seldom observed, and b_2-ions can have a quite high intensity.

The MS/MS identification problem can be formulated as: Given a set of MS/MS spectra $\mathcal{R} = \{r_1, r_2, \ldots, r_n\}$, resulting from the peptides $\mathcal{P} = \{p_1, p_2, \ldots, p_n\}$, generally with known m/z (some of the peptides can be the same), and a set of protein database sequences $\mathcal{D} = \{d_1, d_2, \ldots, d_m\}$, find (identify) the sequences in $\mathcal{D}$ where the peptides $\mathcal{P}$ come from, if any. For each peptide we have three alternatives:

1. The peptide comes from a protein whose sequence is in $\mathcal{D}$.

2. The peptide comes from a protein whose sequence is homologous to a sequence in $\mathcal{D}$.

3. The peptide comes from an unknown protein, with no homologue in $\mathcal{D}$.

Each of the cases above may include *modifications*, the spectrum may therefore come from a peptide that is modified.

In principle, the search is performed by comparing each experimental spectrum to the theoretical peptides (resulting from *in silico* digestion of the database sequences), and identifying the peptides resulting in the best match(es). There may be millions of theoretical peptides in a sequence database however, and it is therefore often impractical to examine every possible peptide. Various types of filtering are therefore used to isolate only those theoretical peptides that constitute potential matches, for example, using the mass of the precursor.

One way of classifying the different approaches for peptide identification is to look at how the *basic comparison* (comparing an experimental spectrum r to a theoretical peptide q) is performed. We divide the methods into three approaches:

Spectral   *compare spectra*: Construct a theoretical spectrum t for the theoretical peptide, and compare the experimental spectrum to the theoretical one.

Sequential   *compare sequences*: Perform *de novo* sequencing of the spectrum to obtain a *derived sequence e*. Then compare the derived sequence to the theoretical peptide.

Threading   *compare spectrum to theoretical peptide*: Either the spectrum is 'threaded' onto the peptide, or vice versa.

The three approaches are illustrated in Figure 4.10. In all approaches, the *scoring* of the matches is essential, and for MS/MS comparison the scoring scheme strongly depends on the matching approach.

The problem can now be formulated as: Given a set of MS/MS spectra $\mathcal{R} = \{r_1, r_2, \ldots, r_n\}$ and a set of theoretical peptides $\mathcal{Q} = \{q_1, q_2, \ldots, q_m\}$, find the theoretical peptide(s) that give the best match(es) to each spectrum.

Such comparisons can take post-translational modifications into account, and also possible mutations (substitutions, insertions, and deletions). Note however that this will make the comparison a lot more complicated.

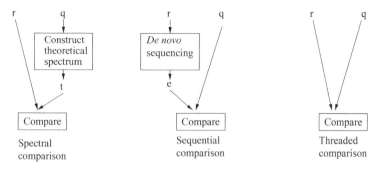

Figure 4.10   Illustration of the three different basic methods for spectrum-to-segment comparison.

4.5.1   Spectral comparison

Spectral comparison means that an experimental spectrum is compared to theoretical spectra constructed from theoretical peptides from a sequence database. The comparison methods can be characterized by the following:

1. which fragment ion types are considered;

2. how the intensities in the theoretical spectrum are calculated;

3. how the comparison is performed (the algorithm) and scored;

4. whether modifications or mutations are taken into account and, if so, how.

4.5.2   Sequential comparison

In sequential comparison possible peptide sequences are *de novo* derived from the spectrum, and these are then searched against the protein sequence database.

The *de novo* peptide sequencing problem via tandem mass spectrometry can be formulated as: *For an experimental spectrum r, a peptide mass m_p, and a set of fragment ion types Δ, derive a sequence (or set of sequences) with mass m_p that gives the best match to spectrum r.*

Post-translational modifications can also be taken into account during *de novo* sequencing. Choosing the best match implies that a score scheme must exist for scoring the derived sequence candidates against the spectrum.

When *de novo* sequencing of a peptide has been performed, it is commonly followed by a sequence database search. However, the spectra are rarely of such a high quality that a unique sequence is derived for the complete peptide. There are also uncertainties in the sequencing due to the similarity of amino acid residue masses, depending on the accuracy of the measurements.

A database search program should be able to deal with such ambiguities and unassigned masses. General sequence database search programs such as BLAST and

FAST have been altered to be able to search with *de novo* derived sequences, but numerous other programs also exists.

4.5.3   Scoring

A similarity score is calculated for each (spectrum, theoretical peptide) comparison. The formula for the score depends on the comparison approach (spectral or sequential) and the method used, but some components are:

- the mass difference between the precursor mass and theoretical peptide mass;
- the mass difference between the fragment masses and theoretical fragment masses;
- the intensities of the peaks identified to be fragments (increasing the score), and of the unidentified peaks (decreasing the score);
- the correlation of the intensities of the fragment and the theoretical expectations of the intensities, calculated from the theoretical peptide.

4.5.4   Statistical significance

The scoring scheme tries to order the segments on the probability that the segments are the origin of the spectrum. It does not, however, say anything about how likely it is that the highest scoring segment really is the origin. For this we should calculate (estimate) a p-value or e-value. The p-value of a score S is the probability of achieving a score of S or higher just by chance in the performed search, and the e-value is the expected number of matches achieving a score of S or higher by chance.

4.6   The protein inference problem

Inferring the proteins in a sample from the detected peptides is complicated by the fact that the mapping from spectra to peptides, and from peptides to proteins is often not unique. Figure 4.11 illustrates different mappings between spectra, peptides, and proteins as part of the identification process:

1. A peptide identified from a spectrum can occur in several proteins, so-called *degenerated or shared peptides* (a, b).

2. Several different peptides can be identified from the same spectrum (c, d), in such cases the spectrum yields a significant score for more than one peptide.

3. A peptide can be identified from several spectra, (d, e).

The points above imply that different sets of proteins can be inferred in a single experiment. When the sets cannot be distinguished using the available evidence, one can employ both the *Occam's razor principle* as well as an anti-razor in an attempt to resolve the issue. The former dictates that one should not invoke novel entities for a

MS/MS spectra

Proteins

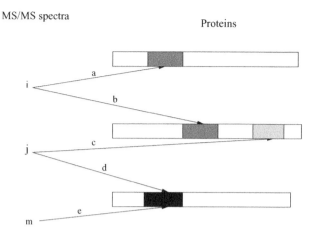

Figure 4.11   Relationship between spectra, peptides, and proteins.

phenomena that can be readily explained without them, that is, the simplest solution is often the correct solution.

Example   Consider the table below with the mappings between the proteins and the identified peptide sequences, assuming that the sequence s_1 is shared between proteins A and D, that s_2 is shared between proteins A and C, and that s_3 represents a unique peptide sequence.

Proteins	Peptide sequences
A	s_1, s_2
C	s_2
D	s_1, s_3

Using Occam's razor will reduce the number of possible sets to two: $\{A, D\}$ and $\{C, D\}$. If additional information is available, for example, that the mapping of s_2 to C requires a modification, and that no additional evidence is found for this, then $\{C, D\}$ is removed from the set, given that $\{A, D\}$ can explain all the observed peptides.

An anti-razor would instead choose the set $\{A, C, D\}$, given that these three proteins could potentially all be in the sample, and because any other set would throw away potential proteins without proper empirical justification.
△

These two views are reflected in what is commonly referred to as *maximal* and *minimal* explanatory sets. The actual protein composition of the sample can be any set that is equal to or broader than the minimal set, and smaller than or equal to the maximal set. Usually, a distinction between equally valid minimal sets (such as between $\{A, D\}$ and $\{C, D\}$ above) is not made, resulting in indistinguishable subsets of proteins in minimal datasets.

4.6.1   Determining maximal explanatory sets

A maximal explanatory set is necessarily dependent on the database used, as well as on the optional search parameters such as allowed missed cleavages and potential modifications. Yet once these have been decided upon, the maximal set can easily be determined by matching each identified sequence against all theoretically derived peptides. For each matching peptide, the precursor protein is added to the set (note that we consider a 'set' to automatically exclude redundant additions). The resulting set composes the maximal explanatory set of proteins for the database used and the search settings employed.

Note that the maximal explanatory set does not necessarily encompass the complete protein composition of the original mixture. Indeed, a substantial number of proteins may have eluded detection altogether. The old adage therefore holds true here, that *absence of evidence is not evidence of absence*.

4.6.2   Determining minimal explanatory sets

Contrary to the maximal explanatory set, construction of a minimal explanatory set of proteins for a given list of peptide masses or peptide sequences is a nontrivial task. Usually a probability is calculated for each matching protein to be present in the sample. These probabilities are typically derived from the confidence one has in the underlying peptide identifications and are mostly designed to work with the results of MS/MS identification algorithms, incorporating for instance the scores attributed to the identifications. It is clear that this strategy will render the construction of a minimal set dependent on external factors such as the underlying search algorithm and any search and/or validation parameters used.

These different algorithms all attempt to solve what is essentially an *ill-posed problem*; there is no way to accurately reconstruct the original protein composition of the sample based on the peptide properties measured. The output delivered by the algorithms therefore necessarily has its limitations and it should be stressed again that follow-up experiments are often the best way to proceed when validation of protein presence (or absence) is required.

4.7   False discovery rate for the identifications

When the number of MS/MS spectra in a search is large it is appropriate to connect the accuracy of the searches to a set of identifications rather than to each identification separately. Given a set of identifications (peptides or proteins) how many of them can assumed to be false? This is the false discovery rate ($FDR = \frac{FP}{FP+TP}$). FDR and similar concepts are described in Chapter 8.4.

Note also that this expression is incorrectly referred to as false positive rate (FPR) in many proteomics papers, and that the concepts can have other meanings in other scientific fields. See also use of FDR and the q-value in Chapter 8.5.2.

One procedure for estimating FDR is as follows:

1. Sort the identifications (peptides or proteins) on the scores.

2. Set a limit S for the scores, considering the matches scoring at least S as correct identifications. These matches constitute the sum of the number of true positives (TP) and false positives (FP).

3. Make a random sequence database, independent of the spectra. This is called a *decoy database*.

4. Perform a new search using the decoy database with exactly the same parameters as in the original search. The number of matches scoring at least S are considered to be equal to FP, since the decoy database should not contain any of the correct sequences.

From this FDR can be estimated, since we have estimates for both FP and $FP + TP$.

4.7.1   Constructing the decoy database

For the procedure above to work the decoy database must satisfy specific constraints. Since it is used to estimate FP for the original search, the search should imitate the search in the original database with the negative spectra (the spectra with no true matches). Thus some properties of the original database should be retained in the decoy database:

• the protein and peptide lengths;

• the amino acid composition, either for each protein or for the whole database;

• the protein and peptide masses.

Several ways of constructing such a decoy database have been proposed, none of which retains all the properties mentioned above.

• Reverse each protein sequence, retaining the protein and peptide lengths, the amino acid compositions and the protein masses, but not the peptide masses.

• Shuffle each sequence, retaining the protein lengths and masses and amino acid compositions.

• Construct random proteins by drawing from an amino acid distribution, either uniform or one corresponding to the distribution in the whole proteome, retaining the protein lengths.

The procedures above are constructed on the protein level. Decoy databases can also be constructed on the (tryptic) peptide level, reversing or shuffling each peptide separately, Bianco et al. (2009). Constructing decoy databases by reversing the sequences is the most common approach, and Bianco et al. (2009) recommend reversing at peptide level.

4.7.2 Separate or composite search

Two different strategies for searching in the original database and in the decoy database are used. One can either perform two separate searches, or make a composite database of the original and the decoy database and perform a single search.

There are no clear recommendations in the literature, but in the mentioned study by Bianco et al. (2009) the separate search approach was found superior on the protein level (calculating FDR from protein identifications). However, no statistically significant difference was found on peptide level (calculating FDR from peptide identifications).

4.8 Exercises

1. In a spectrum there are peaks at $m/z = 821$ and at $m/z = 1231$. Examine if these can come from the same peptide, and determine in that case the peptide mass and the charges.

2. The average amino acid residue mass is approximately 110 Da, and the average number of C atoms per residue is 5.4.
   1. Calculate an isotope intensity distribution for a peptide of mass 800 Da. Consider only the three first isotopes. How does the result compare to Figure 4.5(a)?

   2. Do the same for a peptide of mass 2100 Da.

3. Consider the following values (m/z and intensity) from a raw data spectrum containing a single peak:

842.0095	41.750	842.3329	175.743	842.6564	431.301
842.0389	39.470	842.3624	390.189	842.6858	319.232
842.0683	38.111	842.3918	712.181	842.7153	242.616
842.0977	37.934	842.4212	1078.962	842.7447	181.639
842.1271	39.347	842.4506	1432.524	842.7741	137.138
842.1565	42.621	842.4800	1627.030	842.8035	103.744
842.1859	48.363	842.5094	1542.528	842.8329	78.816
842.2153	58.877	842.5388	1331.617	842.8631	52.775
842.2447	75.060	842.5682	1103.364	842.8928	37.665
842.2741	98.379	842.5976	874.646	842.9212	39.128
842.3035	127.796	842.6270	646.352	842.9321	40.007

   1. Try to determine where the peak starts and stops.

   2. How many raw data points are there for the peak?

4. Consider the following peaks at m/z in an MS/MS spectrum: 72, 129, 175, 276, 299, 331, 363, 395, 418, 519, 565, 622, 675. Can you find a peptide sequence that can produce this spectrum?

   Hint: Extend the peaks with a peak at $m/z =1$, consider the peaks as nodes in a graph, and draw an edge between nodes that corresponds to the mass of one or two residues. Then inspect the paths.

5. Assume that we have identified 7 peptides that together occur in 5 proteins, as shown below.

$$
\begin{array}{c|c}
P_1 & q_1, q_4, q_6 \\
P_2 & q_2, q_6, q_7 \\
P_3 & q_3, q_4, q_5 \\
P_4 & q_3, q_5, q_7 \\
P_5 & q_2, q_6, q_7
\end{array}
$$

   1. Find maximal and minimal explanatory sets.

   2. Assume that the abundances of the peptides are measured as: 600, 400, 200, 600, 200, 400, 600. Can you say which of the minimal explanatory sets are the most probable?

6. Assume that a set of identifications results in the following scores for the matches:

   126.4, 123.9, 117.4, 115.9, 102.3, 98.3, 89.5, 81.2, 77.7, 71.4, 63.5, 58.7, 52.4, 49.7, 46.5.

   Assume further that searches in a decoy database result in the scores:

   99.7, 83.6, 68.2, 51.4, 44.6.

   Calculate the FDR using 7, 10, and 12 of the matches.

4.9   Bibliographic notes

Peak intensity calculation   Garcia and Puimedón (2004), Petrakis (1967), Vaudel et al. (2010).

Spectral normalization   Degroeve et al. (2011), Na and Paek (2006).

Protein inference   Nesvizhskii and Aebersold (2005), Alves et al. (2007), Li et al. (2009).

Decoy databases   Käll et al. (2008), Reidegeld et al. (2008) (program to build decoy databases), Bianco et al. (2009).

5

Protein quantification by mass spectrometry

Protein quantification consists of performing experiments to determine proteome and/or protein profiles and comparing these. Numerous methods are available, and in this chapter we try to classify them and give brief explanations of the different approaches.

5.1 Situations, protein, and peptide variants

Before going into the details of the different approaches we first have to setup the terminology and define some of the most commonly used terms.

5.1.1 Situation

A *situation* is a site, and the state of the site, for which a proteome profile is to be determined (or compared to a profile from another situation). An example of a situation is to look at the liver of cod living in polluted water, or cod living in nonpolluted water.

5.1.2 Protein variants – peptide variants

A quantification task typically compares the abundances of proteins in different situations. To distinguish a protein in different situations/samples we term them *protein variants*, where a different variant is used for each situation or sample (whether the experiment concerns situations or samples depends on the circumstances).

Computational and Statistical Methods for Protein Quantification by Mass Spectrometry, First Edition. Ingvar Eidhammer, Harald Barsnes, Geir Egil Eide and Lennart Martens. © 2013 John Wiley & Sons, Ltd. Published 2013 by John Wiley & Sons, Ltd.

Since the primary sources for quantification information are peptides we accordingly make use of the term *peptide variants* for the occurrences of a peptide in different situations/samples. Note that this does not mean that these peptides or proteins are variants in the strict sense of the word, it only means that they occur in different situations/samples.

5.2   Replicates

The purpose of replicates is to eliminate random variations in order to achieve precise measurements of the subject under consideration, for example, the effect of different treatments for a disease. The idea is that by repeating or replicating the analysis and measurements several times, consistent signals will be confirmed, whereas random variation will be averaged out.

Overall, there are two main types of random variation that can be addressed through replicates:

Biological variation   is natural variation across the biological samples, most often variation between the individuals of the situations under consideration. An example is if we examine the effect of using a specific drug on salmon at age 2. Biological variation can only be addressed by performing replicate analyses, and is often difficult to estimate *a priori*.

Technical variation   is variation in the results due to undesired variations in the experiments, for example, through the use of chemicals from different batches in the sample preparation, instrument drift or random instrument noise in LC-MS/MS runs. Technical variation can be reduced through meticulous attention to a standard operating protocol during sample handling and analysis, but can never be fully eliminated. When experiments are performed at a consistent high standard, the expected amount of technical variation can be predicted relatively well, and tends to be lower than the biological variation.

We thus have two different forms of replicates, each addressing one of the two types of variation.

Biological replicates   that try to cover and average out the biological variation by performing identical experiments on (random) individuals from the situation under consideration.

Technical replicates   that try to cover and average out the technical variation by performing identical experiments on the same individuals (test subjects) or on aliquots of the same sample.

Another term that is occasionally used is:

Experimental replicates   that is often used as synonymous with technical replicates, but in certain cases the replicates are actually classified into three types (biological, experimental, and technical, as in Gan et al. (2007)). Technical replicates are then considered to be replicates within an experiment, whereas experimental replicates are taken across experiments, and biological replicates can occur inside or across experiments.

Note that the distinction between technical and biological replicates is not always clear-cut, and the decision to call a replicate biological or technical is sometimes primarily an issue of semantics.[1]

Results obtained from experiments that employ different replicates can be evaluated by statistical analyses, for example T-statistics or ANOVA, see Chapter 20.5.

As already mentioned, biological variation is often considered the most important, provided that technical variation is controlled through careful experimental setups. Importantly, the number of replicates required to correct for a given amount of variation is proportional to the magnitude of the expected variation compared to the size of the expected effect. If there is a large amount of variation for the measurement of interest in the population, and the effect that needs to be detected on the measurement is small, a large number of biological replicates will be required to lend significance to the observed effect. On the other hand, if the effect is large, and variation is small, far fewer biological replicates can already provide significance for the observed effect.

It is also important to note that there typically are practical limits on the number of replicates that can be run, either due to time constraints, cost factors, or simply the limited availability of samples. In such scenarios, more elaborate experimental designs may be helpful in obtaining significant results despite these practical limitations.

The actual implementation of the experimental strategy when including replicates (number of the different types of replicates required, combination of replicates into single analysis runs, etc.) must be determined in accordance with the quantification method used and the statistical analysis planned.

5.3   Run – experiment – project

A quantification experiment is performed by several mass spectrometry analyses, called *runs*.

5.3.1   LC-MS/MS run

An LC-MS/MS run is described by:

- Input: A peptide sample, or a combination of two or more samples.

- Output: A raw data file containing both chromatographic data and mass spectrometric data.

---

[1] http://bioinformatics.picr.man.ac.uk/mbcf/replicates_ma.jsp.

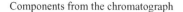

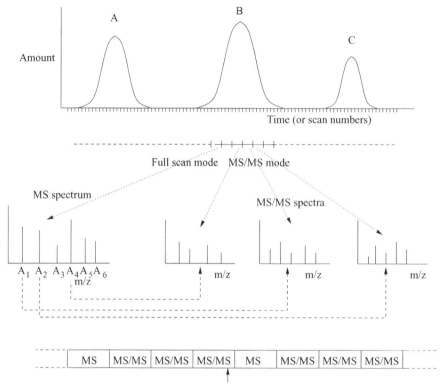

The structure of a file with MS and MS/MS spectra from an MS/MS run.

Figure 5.1   Illustration of the process used for recording LC-MS/MS spectra. Here the three highest peaks from the MS spectrum are selected for MS/MS.

- Process: The sample is subjected to both an LC-process, and a mass spectrometry process, and these two processes are coupled in series but operate in parallel.

Figure 5.1 illustrates an LC-MS/MS run.

5.3.2   Quantification run

A generic quantification run is described by input, output, and the process:

- Input: A set of protein samples, the set can contain one, two, or several samples, depending on the quantification approach.

- Ouput: A set of *peptide tables*, one for each LC-MS/MS run. The contents of the peptide tables are different for the different quantification approaches, but typically contain mass and intensities of peaks/peptides, and can also include retention time. Peptide tables are further discussed in Section 5.10. More detailed description of specific input and output will be provided in the chapters detailing the different approaches.

- Process: Typically consisting of the following:
    - digest and possibly label each sample, producing peptide sample(s);
    - combine the peptide samples if more than one sample;
    - fractionate the peptide sample, if it is too complex for one LC-MS/MS run;
    - perform an LC-MS/MS run for each *fraction*;
    - *in silico* analyze the data obtained in each MS/MS run, making a peptide table.

When the number of samples that can be compared in a single quantification run is greater than two we term it *multiplexing* in analogy to the mixing of different signals on a single channel in telecommunications and computer networks.

5.3.3   Quantification experiment

A model of a generic quantification experiment is illustrated in Figure 5.2.
   A quantification experiment can be described by:

- A specific set of situations, often two, that are to be compared.

- For each situation a number of biological and technical replicates are used. The replicates are chosen in accordance with the quantification method and the statistics to be used.

- A set of quantification runs to obtain the measurements.

- A statistical analysis based on the resulting peptide tables, producing a final quantification report.

5.3.4   Quantification project

The distinction between project and experiment can seem somewhat arbitrary, and in this book we focus on experiments. For projects we only state that they can consist of several experiments that can be of different types.

5.3.5   Planning quantification experiments

It is important to first define the question(s) to be answered, and the methodology to be used, before performing a quantification experiment. The following list contains a summary of the actual tasks involved in setting up an experiment (not necessarily

A general model of a quantification experiment for n situations

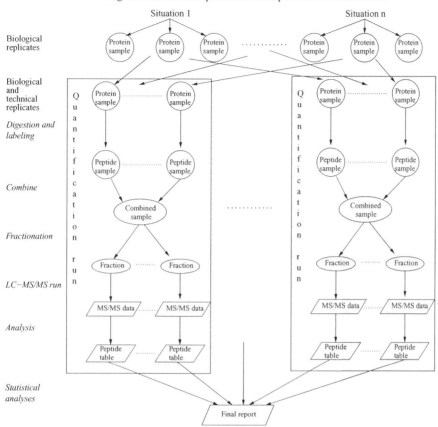

Figure 5.2   Illustration of a protein quantification experiment using LC-MS/MS.

in the given order):

- Determine the *situations* to be used in the experiment.

- Determine the main analysis task, in protein quantification this is often the detection and identification of proteins that are differentially abundant between the considered situations.

- Determine the sample types, the use of *pooled* samples (several individuals combined in a single sample) or single, *individual* samples.

- Determine the number of replicates to use.

- Determine the type of comparison: Paired or non-paired. In a paired comparison, pairs of individuals are made from the samples, for example, a given individual before and after taking a drug.

- Determine the statistical analysis methods to use.

- Determine the experimental tools to use, and their operational parameters:

  - which instruments to use, for example, type of LC-MS/MS equipment;

  - which software to use for analyzing the produced data.

- Determine the quantification runs to perform, and decide on the distribution of the different samples (replicates) over the planned quantification runs (experimental design randomization).

One challenge in quantitative proteomics is to perform the whole process (sample gathering and preparation, mass measurement, etc.) such that the resulting outcome from the different samples can be compared. This usually requires the situations/samples to be 'similar,' for example, having approximately the same total protein content, and that the abundances of most proteins are the same across the samples.

Other challenges are the previously mentioned large dynamic range of protein abundances and the difficulty of achieving reproducibility when employing complex instrumentation to analyze highly complex samples.

5.4   Comparing quantification approaches/methods

A quantification experiment consists of a number of steps, and the reliability of the results depends on the quality of all the steps. In addition to the sample processing and the instruments used, for the computational/statistical aspect it depends on:

- the calculation of the peak intensities;

- the determination of the corresponding abundances across samples/situations;

- the peptide identification;

- the protein inference.

Below we describe some challenges related to characterizing and comparing different quantification approaches.

5.4.1   Accuracy

Accuracy is a measure of the closeness of a result to the true value. It involves a combination of a random error component and a common systematic error or *bias* component. It depends on several factors in the experiment, but one of the most critical is the peak detection in the spectra.

5.4.2   Precision

Precision is the closeness of agreement between independent measurements obtained by applying the experimental procedure under stipulated conditions. The smaller the random part of the experimental errors which affect the results, the more precise the procedure. A measure of precision is the standard deviation, or a formula based on the standard deviation.

The value of a measurement can be precise but not accurate, neither of them, or both of them. It is not possible to achieve high accuracy without high precision; if the values are not similar to one another, they cannot all be similar to the correct value. However it is possible to achieve high average accuracy without high precision. If the value from a measurement is both accurate and precise, it is often said to be *valid*.

Example Let the correct mass of a peptide be 950.455 Da, and let two instruments perform four measurements each, resulting in the following value sets: {950.650, 950.654, 950.652, 950.653}, and {950.465, 950.485, 950.445, 950.425}. The first instrument has high precision, but low accuracy; while the other has higher accuracy, but lower precision.

△

Accuracy and precision are illustrated in Figure 5.3. The bias (or the systematic error) of the measurement depends on several 'fixed' factors, and is usually defined (for the instrument and method used) as the mean of the deviations from the correct value. It is often used as a measurement of the accuracy of the instrument, and can also be used for calibration purposes. In fact, the use of the mean value of multiple measurements of a standard for calibration yields a two-step procedure that attempts to minimize the random fluctuations in the instrument measurements while compensating for the systematic error.

Precision is often determined by taking a sufficient number of measurements of the same sample, and performing a statistical analysis on the data. This analysis most

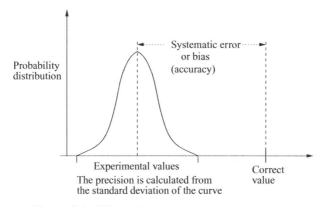

Figure 5.3   Illustration of accuracy and precision.

often assumes a normal distribution for the measurement errors and relies on the 95 % confidence interval, calculated as 1.96 times the standard deviation (for a two-sided test against a normal distribution). In many cases however, it is not convenient or even not possible to perform multiple measurements, and the precision then needs to be estimated indirectly, with the exact mechanism depending on the type of attribute measured as well as the instrument used.

Note that the dynamic range (see below) and complexity of the sample has a great impact on the accuracy and precision.

5.4.3   Repeatability and reproducibility

When a more precise definition of precision is needed, repeatability or reproducibility is used. Repeatability considers the same method with the same test material under the same conditions (same operator, same apparatus, same laboratory, and after short intervals of time) whereas reproducibility means the ability to repeat someone else's experiment using the same test material, but under different conditions (different time, laboratories, instruments, etc.) and achieve the same results.

5.4.4   Dynamic range and linear dynamic range

The dynamic range of a measurement (or detector) is the range between the lowest and highest values that can be measured. Most often it is the linear part of the dynamic range that is interesting. Linear dynamic range means that if the real abundances of two components (or the same component in different situations) are a_1 and a_2 respectively, and they are measured to x_1 and x_2 respectively, then

$$\frac{x_2 - x_1}{a_2 - a_1} = k$$

for a constant k. The linear dynamic range is the interval for which this equation yields.

5.4.5   Limit of blank – LOB

EP17[2] defines LOB as *the highest apparent analyte concentration expected to be found when replicates of a sample containing no analyte are tested.* It is proposed to be estimated by replicative measuring of a sample containing no analytes, and defining

$$\text{LOB} = \mu_b + 1.645\sigma_b,$$

where μ_b, σ_b are the mean and standard deviation of the measures respectively. Assuming Gaussian distribution LOB represents 95 % of the measurements.

---

[2] Clinical and Laboratory Standards Institute. Protocols for Determination of Limits of Detection and Limits of Quantization, Approved Guideline. CLSI document EP17. Wayne, PA USA: CLSI; 2004.

5.4.6   Limit of detection – LOD

Limit of detection (LOD) of a method or instrument is the minimum quantity of a component that can be confidently detected (can be reliable distinguished from LOB). It is also termed *level of detection*. It is often estimated by replicative measuring of a sample known to contain a *low abundance av analyte*, and calculated as

$$LOD = LOB + 1.645\rho_s,$$

where ρ_s is the standard deviation of the measures. Again assuming Gaussian distribution, 95 % of the values will be greater than LOB. Other ways of estimation have been proposed, especially when a Gaussian distribution cannot be assured, see Bibliographic notes.

5.4.7   Limit of quantification – LOQ

Limit of quantification (LOQ) of a method is the minimum amount of a component that can be confidently quantified. It should be greater than LOD, and for mass spectrometry is often estimated as 3 x LOD.

5.4.8   Sensitivity

When talking about sensitivity with regards to mass spectrometry we mean the ability to detect and quantify signals from low abundant components, since this is the most used understanding of the term in the field of proteomics. It can be quantified by limit of detection or limit of quantification. High sensitivity means low LOD.

However, be aware that for mass spectrometry, sensitivity is formally defined as the ratio of the change in ion current to the change in the sample.

5.4.9   Selectivity

Selectivity means a method's ability to distinguish and quantify a particular component (peptide) in a mixture without interference from other components. This has considerable impact on the reliability of the results.

5.5   Classification of approaches for quantification using LC-MS/MS

Mass spectrometry quantification experiments can roughly be divided into two main tasks: *Identification* and *quantification*. Identification means determining the peptide and inferred protein origin of an MS/MS spectrum, and is typically achieved by searching the acquired MS/MS spectra with specialized software against protein sequence databases (see Chapter 4). Quantification on the other hand, can be based on information from either the MS spectra, or the MS/MS spectra.

Table 5.1 Classification of the different LC-MS/MS approaches.

	Type	Labeling	Abundance determination
1	Discovery	Label based MS/MS quantification	Ion current
2	Discovery	Label based MS quantification	Ion current
3	Discovery	Label free	Ion current
4	Discovery	Label free	Peptide identification
5	Targeted	Label based	Ion current
6	Targeted	Label free	Ion current

The order in which identification and quantification are performed depends on the method used. In practice both are performed at the peptide level in the situations, and protein quantities and identities are inferred from the peptide-level results.

In order to classify the different approaches (or strategies) for quantitative proteomics, we will now consider these approaches from different viewpoints, as elaborated below (see also Table 5.1).

5.5.1 Discovery or targeted protein quantification

A simple way of categorizing quantification methods is to consider whether the proteins of interest (to be quantified) are unknown or known *a priori*. In the former case the unknown proteins of interest, for example, those that are differentially abundant, have to be *discovered*, whereas in the latter case, the known protein(s) of interest can be *targeted*.

Discovery quantification   This approach is mainly used for recognizing proteins that are differentially abundant in two or more situations. It relies on which peptides are selected for fragmentation (MS/MS processing). Given that this selection contains some randomness the approach is not (fully) reproducible. Identification of the *a priori* unknown peptides is performed using standard database searching with MS/MS spectra, and the unknown proteins inferred. Methods using protein discovery quantification are described in Chapters 9-14. Discovery quantification is mainly performed by shotgun proteomics, and therefore also commonly termed *shotgun quantification*. We prefer discovery as it is a better term to distinguish it from targeted. Note also that shotgun can be principally used when the proteins in the sample are known.

Targeted quantification   In targeted quantification the proteins to be quantified are known. Thus it is an appropriate approach for testing biological hypotheses, for example, in systems biology. The number of proteins in the sample(s) can be large, but the number of proteins quantified is limited, typically some hundreds. Targeted quantification is mainly performed by a method called *Selected Reaction Monitoring (SRM)*. Use of SRM typically results in higher sensitivity, and better signal-to-noise-ratio than for discovery quantification. SRM is described in Chapter 15.

5.5.2   Label based vs. label free quantification

In peptide quantification we have to be able to distinguish between the variants of a specific peptide that come from different situations or different samples. This can be done by tagging the peptides with different labels depending on the situations/ samples, that is, label based quantification, or by treating the different situations/samples in different LC-MS/MS runs, and simply keeping track of which situation/sample was analyzed in which run that is, label free quantification.

Note that we use the combined term situations/samples. This is because there will generally be several samples for a situation, and in some cases it is more appropriate to consider variants associated with situations, and in other cases variants associated with samples.

The labels are often referred to as *reagents*, and we will use these two terms interchangeably.

Label based   The proteins from different situations/samples are labeled differently, such that peptide variants can be distinguished either in the MS spectra or in the MS/MS spectra. Peptides from the different situations are mixed, and the LC-MS/MS runs are performed on the mixed sample. This means that the variants of a peptide (or fragment) occur in the same spectra, but with different masses. Label based methods are described in Chapters 9-12.

Label free   LC-MS/MS runs are performed separately for the peptide samples from each situation. This means that the peptide variants occur in spectra from different LC-MS/MS runs, with approximately the same masses and retention times, depending on instrument accuracy, precision, calibration, and drift. Label-free methods are described in Chapters 13 and 14.

Comparison of label based and label free

- The label free methods are less costly and biologically simpler than the label based ones since no additional sample preparation (labeling) is required.

- The corresponding peptide variants are more difficult to detect in label free methods than in label based, since in label based they are in the same spectra, but in label free they are in spectra from different runs. Thus high precision and accuracy is vital, and high resolution mass spectrometers needed. In addition, developing reliable programs for data processing is more challenging for the label-free methods.

- The label-free methods support higher dynamic ranges compared to label based, typically 1:60 compared to 1:20 for label based.

5.5.3   Abundance determination – ion current vs. peptide identification

The peptide abundances (and hence the protein abundances) can generally be determined in two different ways:

Ion current   The intensities of the peaks corresponding to the peptides (and thus to the inferred proteins) are used. The peaks considered can either be the precursor peaks in the MS spectra (the most common approach), or fragment ion peaks in the MS/MS spectra. This approach is described in Chapters 9-13.

Peptide identifications   Calculation of the abundance of a protein is based on the number of times a peptide from the protein is selected for MS/MS processing. There are several variations depending on the exact metrics used. This approach is described in Chapter 14.

Comparison of ion current and peptide identification

- Ion current based quantification has been shown to be more accurate and to support a larger dynamic range.

- Ion current can also be used for unidentified peptides, thus one can first recognize peptides that are differentially abundant, and then identify the peptides and infer the proteins.

- Use of peptide identification requires less complicated software.

5.5.4   Classification

The above definitions make it possible to classify the methods as shown in Table 5.1. Note that for Discovery – Label based we distinguish between MS quantification and MS/MS quantification, depending on which type of spectra are used for quantification.

5.6   The peptide (occurrence) space

The protein abundances are calculated from the abundances of the peptide variants. We should therefore know where abundances of a specific peptide variant, p_v, can be found, that is, where peaks from the peptide variant can be found. This is called the *peptide space*. The recognition of the peptide space varies with the quantification approaches used. By considering Figures 5.1 and 5.2 the following should be taken into account:

1. different copies of p_v can fall into different fractions;

2. for one LC-MS/MS run different copies of p_v can (occasionally) occur at different chromatographic peaks (eluting at different time intervals);

3. for one chromatographic peak there will usually be several MS spectra, such that p_v can occur in several (successive) MS spectra;

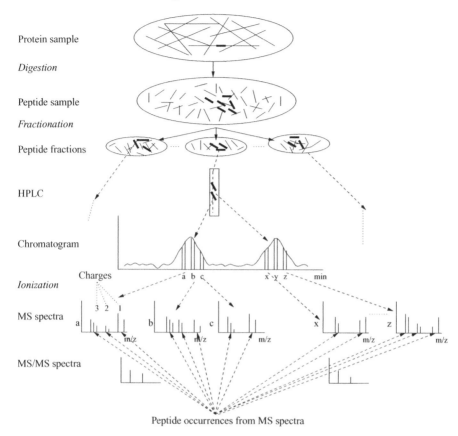

Protein sample

Digestion

Peptide sample

Fractionation

Peptide fractions

HPLC

Chromatogram

Ionization

MS spectra

MS/MS spectra

Peptide occurrences from MS spectra

Figure 5.4   Illustration of the peptide space of a peptide variant. Copies of a specific peptide variant are highlighted.

4. if the peptide is multiply charged, different copies of p_v can appear in the same spectrum with different charge states and therefore different m/z ratios;

5. p_v will normally occur as an isotope pattern.

This is illustrated in Figure 5.4, where charge states from one to three are shown. Note that for a specific run only a subspace of a peptide space is found and used in the quantification.

Comparing the abundances of two or more variants of a peptide first and foremost requires the recognition of the corresponding peaks. Finding such corresponding peaks is not self-evident however, and the complexity of this task varies with the approach. Corresponding peaks can be in the same spectrum, or can be in different spectra, primarily depending on whether labeling is used. One must also keep in mind that a peak of a peptide variant can interfere with other peptides, such that the observed abundance is not derived from the peptide variant under consideration alone. This can be a serious problem, and is called *peptide interference*.

The procedures for recognizing corresponding peaks are therefore described under the different approaches, where the connection between peak intensities and peptide abundances are also explained. Note however that the total occurrence space of a peptide variant will not necessarily be recognized, and that often corresponding sub-occurrence spaces are used in the calculation of relative protein abundances instead.

5.7   Ion chromatograms

In many quantification methods peak intensities from successive spectra are combined, and the combined intensities are shown as chromatograms, often called *ion chromatograms*. There are various types of ion chromatograms:

Total Ion Chromatogram – TIC   shows the total intensity from all m/z values that are detected in successive spectra.

Selected Ion Chromatogram – SIC   shows the ion intensity inside a small m/z-window along the retention time. Here only signal from components in the selected m/z-window is transmitted and/or detected by the MS-instrument, the scanning is performed in selected ion mode (*Selected Ion Monitoring – SIM*). It is also termed **Single Ion Chromatogram**.

eXtracted Ion Chromatogram – XIC   has the same appearance as a SIC, but it is reconstructed from ions produced over wider m/z windows. Figure 5.5 illustrates an XIC constructed from several spectra.

Given that an XIC is reconstructed from a larger dataset than a SIC, the quality of an XIC appears to be lower than that of a SIC. However, an XIC is often used to construct chromatograms yielding a single isotope where the m/z values can vary over the larger set of spectra, as shown in Figure 5.5. An XIC can also be constructed for an isotope pattern or a charge pattern, thus combining several m/z windows. In this way relative peptide abundance can be calculated by comparing the XIC of two variants of the peptide.

5.8   From peptides to protein abundances

The procedure used to go from spectra to peptides to protein abundances varies considerably with the quantification approaches, from the simplest case where protein abundance is calculated directly from spectra, to the more advanced cases where comprehensive peptide abundances are first calculated from spectra prior to inferring protein abundances. In order to minimize the number of terms employed, we will here use the term peptide abundance, even when spectrum abundance is strictly the most correct term.

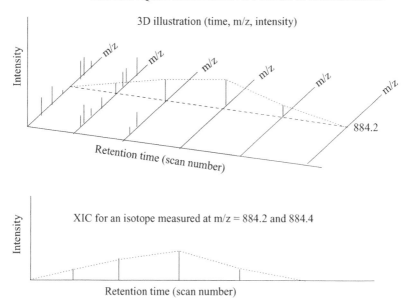

Figure 5.5 An XIC for an isotope with measured m/z of 884.2 in three of the scans, and 884.4 in one. For clarity, not all peaks are shown in the spectra.

In order to provide an overall model of the different methods/approaches for protein quantification, we first detail the terms used.

Elementary abundance is a measured abundance that is supposed to come from a single peptide variant. It can be the intensity of a single monoisotopic peak, or the intensity of an entire isotope envelope, or other collections depending on the approach. Protein abundances are calculated from a set of such elementary abundances.

Single abundance is abundance of a peptide or a protein in one situation (sample). Elementary abundance is an example of an single abundance. Single is used here in contrast to relative abundance, but should not be confused with absolute abundance.

Relative abundance is the abundance of a peptide or a protein in one sample or situation divided by the corresponding abundance in another sample/situation. The resulting number thus becomes a fraction without unit.

In what follows, we consider two samples. Let $\{a_1, \ldots, a_n\}$ be the elementary abundances from sample 1, and let $\{b_1, \ldots, b_n\}$ be the corresponding abundances from sample 2, see Figure 5.6 where corresponding elementary abundances occur in the same MS spectrum.

A single peptide abundance can be a combined abundance $a = f(a_1, \ldots, a_n)$, where f is a function showing how a is calculated from the elementary intensities.

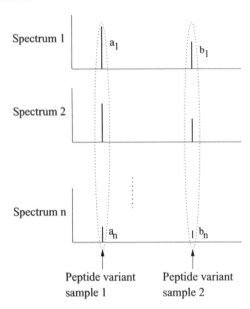

Figure 5.6 Illustration of corresponding elementary abundances $(a_1, b_1, \ldots, a_n, b_n)$.

A relative peptide abundance can be a combined abundance $r = f^*(a_1, \ldots, a_n, b_1, \ldots, b_n)$, where f^* is a function showing how r is calculated from the elementary intensities.

We now consider the calculation of relative protein abundances from elementary abundances. This means that the calculation derives relative abundances from single abundances. How early in the calculation this happens depends on the specific quantification method used. However, the quantification generally consists of three central tasks:

1. from single abundances to combined single abundances;

2. from corresponding single abundances to relative abundances;

3. from relative abundances to combined relative abundances.

Note that the results can yield both peptide or protein abundances. The three steps will be explained in more detail below.

5.8.1   Combined single abundance from single abundances

The combined abundance can either be the sum of the single abundances, or the average, depending on the circumstances. Average abundance can for example be calculated as the mean, through weighting, or as the median, as shown below.

Mean

$$a = \frac{1}{n} \sum_{i=1}^{n} a_i$$

Weighted mean

$$a = \frac{1}{n} \sum_{i=1}^{n} w_i a_i,$$

where w_i is a weight, for example calculated from uncertainties (errors or signal-to-noise ratios) in the peptide abundance determinations.

Median   of the a_is, making the average robust to outliers.

5.8.2 Relative abundance from single abundances

The most common methods for calculating relative abundance r are:

Ratio of sums

$$r = \frac{\sum_{i=1}^{n} a_i}{\sum_{i=1}^{n} b_i}.$$

Linear regression   is used to fit a line through the points $\{a_i, b_i\}$. This means that one considers errors in only one of the measurements (either in the $\{a_i\}$'s or in the $\{b_i\}$'s). The slope of the line is used as the value for r.

Linear correlation   is used to fit a line, meaning that errors in both measurements are taken into consideration. Again the slope is used.

An example of a linear relation is shown in Figure 5.7(b).

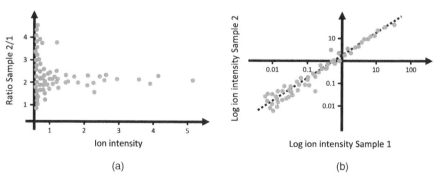

(a)   (b)

Figure 5.7   Example of linear relations for relative intensities (a), and log transformed intensities (b).

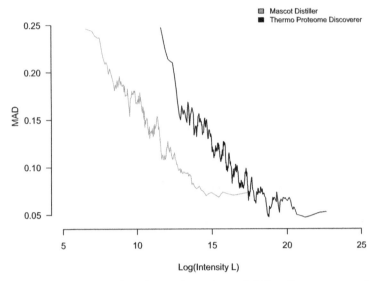

Figure 5.8   Plot showing that the increase in variability (here measured as the Median Absolute Deviation or MAD) for the ratio is higher for low intense peptides. Lines are plotted for two distinct quantification algorithms, applied to the same dataset.

Correlation and regression are described in detail in Chapters 20.8 and 20.9. Correlation is most often preferred, but Carrillo et al. (2010) investigated several of the methods, and found the 'Ratio of sums' method to be the best estimator for the true protein abundance.

It is well known that the variability in abundance is highest for low abundance analytes (Colaert et al. (2011b)), as illustrated in Figures 5.7 and 5.8. As a consequence of this the lowest abundant peptides are often either discarded before protein abundance is calculated, or a variance stabilization method is used to equalize the variances across the abundances, see Chapter 6.6.5.

5.8.3   Combined relative abundance from relative abundances

Let $r_1, \ldots, r_n$ be relative abundances. The combined relative abundance cr, can then be calculated as:

Mean

$$cr = \frac{1}{n} \sum_{i=1}^{n} r_i$$

Weighted mean

$$cr = \frac{1}{n} \sum_{i=1}^{n} w_i r_i,$$

where w_i is a weight typically determined from a_i, b_i, giving highest weights to highest abundances, but uncertainties in the identification or abundance determination (signal-to-noise ratios) are also used.

Median   of the r_i's, making the average robust to outliers.

5.9 Protein inference and protein abundance calculation

In Chapter 4.6 we discussed the protein inference problem related to protein identification. Protein inference also affects protein quantification and is in fact one of the most challenging tasks.

5.9.1 Use of the peptides in protein abundance calculation

Below we briefly discuss the most serious issues related to which peptides should be used in calculating the protein abundances.

An identified peptide occurs in more than one protein (degenerate or shared peptide)

In this case one cannot directly know which of the possible proteins are in the sample(s), and in which quantities. The easiest way of dealing with these peptides is to not use them in the quantification, that is, only use uniquely identified peptides (*unique peptides*) when calculating the protein abundances. However, a consequence of this is that it is difficult to quantify proteins not containing unique peptides, and often impossible.

Several peptides are identified from a spectrum

This means that the tool used for identification returns several possible peptides with scores above the significance level. The easiest and most common way of dealing with this is to use only the most significant peptide, that is, the highest scoring one. This will mainly be the case for our treatment of this subject. However, more comprehensive methods can take into account several high-scoring peptides throughout the analysis workflow.

A peptide is identified by several spectra

The exact way of treating this in the protein abundance calculation depends on the quantification method. Often it is simply treated either as if two different peptides have been quantified, or an average abundance is calculated across the spectra for the peptide.

A peptide is identified with low significance

Usually there is a threshold for filtering out spectra identified with low significance.

One can also consider removing spectra from use in the abundance calculation due to other reasons, for example, due to low total intensity, poor signal-to-noise ratio, or due to some peaks reaching the intensity saturation limit for the mass instrument used.

Also one should test for outliers among the possible peptides of a protein.

5.9.2   Classifying the proteins

In order to determine which proteins can be quantified one tries to classify the proteins. A common classification is:

Unique protein: A protein for which there exists (at least) one unique peptide.

Protein group: A group of proteins which contain exactly the same set of identified peptides.

Subset protein: A protein for which all its identified peptides are contained in another protein.

Shared protein: A protein with no unique peptides, but is not a subset protein.

The proteins in a protein group are indistinguishable, and usually one protein is selected to represent the whole group. Also note that both the subset proteins and the shared proteins are difficult to quantify.

5.9.3   Can shared peptides be used for quantification?

When discussing the use of shared peptides two issues have to be considered:

- can the use of the shared peptides increase the reliability of the quantification?

- can proteins with only shared peptides be quantified?

Dost et al. (2009) proposes a method for this where a set of equations are developed. For a brief description of the method consider:

- two protein samples S, T, for which we will find relative abundances;

- a group of proteins $P_1, P_2, \ldots, P_n$;

- a group of peptides $q_1, q_2, \ldots, q_m$.

The groups are formed such that:

- the peptide group is formed of all the peptides identified for the proteins in the group;

- each protein shares at least one peptide with at least one other protein in the group;

- none of the peptides occur in proteins not in the group.

Example Let

- Protein P_1 has the identified peptides q_1, q_2, q_3;

- Protein P_2 has the identified peptides q_2, q_4;

- Protein P_3 has the identified peptides q_3, q_4, q_5.

$\triangle$

Let

- A_i^S, A_i^T be the (single) abundances of protein P_i in the samples, respectively;

- r_j be the (known) relative abundance of peptide q_j in the two samples.

Assume that a peptide q_j is shared between the proteins P_1, P_2, P_3. An equation is then formed:

$$\frac{A_1^S + A_2^S + A_3^S}{A_1^T + A_2^T + A_3^T} = r_j, \tag{5.1}$$

with $\{A_i^X\}$ being the unknowns, $X = S, T$.

Similar equations are formed for each of the peptides. Hence the number of equations is equal to the number of peptides, m. For n proteins there are $2n$ unknowns. Since the unit for $\frac{A_i^S}{A_i^T}$ is immaterial an additional equation can be formed by:

$$\sum_{i=1}^{n} A_i^S = k \text{ (k is a constant, for example 100)}.$$

Thus we have a linear system of $m + 1$ equations and $2n$ unknowns. Whether this is solvable depends on the ratio of $m + 1$ to $2n$ and on whether the equations are linearly dependent.

In this way the relative abundance of proteins containing only shared peptides can be estimated if the set of equations is solvable.

Example Considering the previous example there will be six unknowns and six equations, and a unique solution exists if the set of equations is not linearly dependent. Thus protein P_2 containing only shared peptides is quantified.

$\triangle$

A linear programming problem for the general case can be formed by introducing an error function and minimizing the total error, and it is also possible to take into account a *detectability score* for the peptides. The detectability score says something about how probable it is that the peptide should be identified, mainly based on the probability of being ionized.

Additional proposals for handling shared peptides specialized for spectral counts are described in Chapter 14.

5.10   Peptide tables

As mentioned earlier each identified MS/MS spectrum usually comes with an associated peptide abundance. The abundance can be calculated from peaks in the considered MS/MS spectrum, or from a peak in the precursor MS spectrum from which the MS/MS spectrum is selected.

It is often appropriate to collect such (temporary) calculations in a peptide table. A peptide table contains one entry for each MS/MS spectrum. Given that the different quantification approaches vary so much, the content of the peptide tables also varies, and depends on:

- how many peptide variants an MS/MS spectrum can represent; (iTRAQ: four or eight, SILAC: two or three, unlabeled: one);

- how the peptide variants are determined from the spectra;

- which type of peptide abundance is obtained, relative or single;

- how the abundances are calculated, and from which type of spectra (MS or MS/MS);

- which type of database(s) are used for identifications;

- how the protein abundances are calculated from the peptide abundances.

5.11   Assumptions for relative quantification

As discussed above, relative quantification considers only the change in abundance of the peptide or protein variants across the situations, rather than as in absolute quantification where the copy numbers are wanted. Depending on the specific method employed, relative quantification can often be a lot easier and more straightforward to calculate than absolute quantification. It is important to realize however, that the existing approaches for relative quantification are all based on the following assumptions:

- The abundance of the large majority of the proteins do not change across the considered situations/samples. This allows us to establish a baseline for the accuracy and precision of the relative quantification, since we expect most proteins to have a ratio of 1/1.

- The change in abundance of a subset of the proteins will have a negligible effect on the quantification and normalization process. This is important since we expect outliers (significantly changed proteins) to be present in the sample, or it would be useless to go looking for them. Yet the methods employed to analyze the data are not always robust against outliers, so it is important to take this into account when processing the data. Outliers are described in Chapter 6.4.

- The linear range of the detector response is broad enough to encompass the ratios of the analytes across the considered situations/samples. If the detector response starts to flatten out, and it no longer responds linearly, for example, if there is a lot more of the peptide or protein in one of the situations as compared to the other, relative quantification simply becomes *differential analysis*: While we can still say that a protein variant is more highly abundant in one situation/sample, we can no longer ascertain how much more abundant it is.

5.12   Analysis for differentially abundant proteins

We can now consider two main procedures for the path from elementary abundances to relative protein abundances and statistical analysis.

Procedure *A*

1. calculate single protein abundances from the elementary abundances for each sample;

2. compare all the single protein abundances to find differentially abundant proteins, for example, by *T*-statistics or ANOVA.

Procedure *B*

1. calculate relative peptide abundances from the elementary abundances, using pairs of corresponding samples;

2. calculate relative protein abundances from the relative peptide abundances;

3. analyze the relative protein abundances for differentially abundant proteins, for example, by fitting all relative protein abundances to a normal distribution, and looking for outliers.

5.13   Normalization of data

Normalization of data in the context of proteomics data analysis can have two different meanings:

1. the transformation of a dataset to a normal-like distribution, such that parametric test statistics can be used; or

2. the removal of shifts or noise in the data resulting from systematic or random shifts in instrument performance and measurement, such that data from different runs can be compared.

In order to distinguish these two interpretations of the term normalization, we refer to them as *statistical normalization* and *experimental normalization*, respectively. However we will occasionally use the term normalization alone when the implied meaning is obvious from the context. Statistical normalization is described in Chapter 6, experimental normalization in Chapter 7.

5.14   Exercises

1. Assume that we are to compare two samples, and label one of the samples with a label of mass 12 Da. We now consider a peptide of mass 948 Da for which there is only one labeling site. Complete labeling (all individual peptides labeled) is assumed. The peptides may have one, two, or three charges. Assume that the peaks of an MS spectrum are (m/z, intensity):

   (312, 2340) (321, 617) (365, 8443) (412, 6534) (436, 10 346) (475, 6334) (481, 9546) (512, 5678) (567, 18 253) (613, 9312) (666, 12 376) (714, 5349) (794, 8573) (812, 2643) (867, 3912) (912, 3567) (949, 9612) (961, 13 247) (988, 3465).

   Try to find peaks corresponding to the two versions of the peptide, and determine a value for the relative abundance between them.

2. Let the mass of a peptide be 420.38 Da. Assume that the peptide occurs in a set of succeeding MS spectra and is measured as 419.60, 419.72, 419.78, 419.94, 420.08, 420.12, 420.26, 420.36 Da.

   • What is the average accuracy?

   • Calculate a value for the precision.

3. We are going to test a new procedure for determining the relative abundances of proteins. We consider a specific protein P in a sample for which there is an abundance (concentration) of 50 $fmol/\mu l$. We make a labeled version of the protein, which is spiked in with different abundances. For each abundance value we perform an LC-MS run, and get the following values:

Abundance spiked in	0.1	0.5	1.0	5.0	10.0
Measured relative to 50	0.001	0.008	0.022	0.098	0.218

Abundance spiked in	50.0	100.0	500.0	1000.0
Measured relative to 50	0.976	1.964	10.048	14.012

   We now want to determine a possible linear dynamic range. As a graphical illustration we will use a scatter plot, which means that the horizontal axis will represent the spiked-in abundances, and the vertical axis will represent the measured relative abundances. Given that the values have a large range we will take the logarithm of the values before plotting. Make such a plot an try to determine a linear dynamic range.

4. Let the limit of a blank run of an instrument be 40. Assume that we have a low abundance (20) of a protein, and perform several experiments, measuring the

abundances to be 18.2, 18.6, 19.1, 19.6, 20.2, 20.5, 20.6. Calculate a value for LOD and LOQ.

5. Assume that we have an LC-MS/MS run, and consider a specific peptide. Let the peptide elute in a time interval of 2 min, and an MS spectrum is made every 3 sec. Assume further that the peptide can have a charge of 1, 2, or 3. How large is the peptide space (the maximum number of peaks corresponding to the peptide)? (Consider a possible isotope pattern as one peak.)

6. We want to calculate the relative abundance of a protein between two samples, and have identified four peptides belonging to the protein. For each peptide we have found the following corresponding peaks with the respective intensities:

Peptide 1	(2040, 640) (1400, 495) (4230, 1430)
Peptide 2	(4270, 1020) (2340, 605) (1070, 245)
Peptide 3	(820, 295) (3130, 1050) (2400, 1210) (1860, 640)
Peptide 4	(6210, 2110) (3040, 1020)

Calculate a value for the relative abundance of the protein. Consider if some of the measurements or derived values can be outliers (deviating largely from other values), and thus should not be included.

7. We have the following combination of identified peptides and possible proteins. Classify the proteins into the following categories: Unique protein, protein group, subset protein, and shared protein.

P_1	q_1, q_2, q_3, q_4
P_2	q_2, q_5, q_6, q_7
P_3	q_3, q_4
P_4	q_2, q_5, q_6, q_7
P_5	q_2, q_5

8. Consider the first example in Section 5.9.3. Let the relative peptide abundances be $r_1 = 2.0, r_2 = 1.43, r_3 = 1.14, r_4 = 0.75, r_5 = 0.5$. Try to determine the relative abundance of the shared protein P_2. *For the additional equation you can use* $A_1^S + A_2^S + A_3^S = 24$.

9. We have the following (single) values for protein abundances in two comparable samples:

Sample 1	8	12	9	17	6	13	4	16	9	6	11	12
Sample 2	13	18	28	25	8	20	6	14	14	9	16	17

Sample 1	3	18	7	15	17	9	5	11
Sample 2	5	12	10	22	25	14	8	16

We now want to find differentially abundant proteins. We employ the assumption that only a small number of the proteins are differentially abundant. Perform an experimental normalization in accordance with the assumption such that the abundances from the two runs can be compared. Then indicate which proteins seems to be differential abundant. *Hint: You can, for example, assume equal total abundances.*

5.15   Bibliographic notes

Reviews   Vaudel et al. (2010), Bantscheff et al. (2007b), Mueller et al. (2008), Panchaud et al. (2008).

Limit of blank, detection, quantification   Currie (1968), Linnet and Kondratovich (2004), Armbruster and Pry (2008), Keshishian et al. (2009).

Accuracy and precision   Karp et al. (2010).

From peptides (spectra) to protein   Carrillo et al. (2010), Bantscheff et al. (2007b), Nesvizhskii and Aebersold (2005).

Comparing different approaches   Vaudel et al. (2010), Domon and Aebersold (2010).

6

Statistical normalization

In Chapter 5.13 we mentioned that normalization of data in the context of proteomics can have two different meanings: Statistical and experimental normalization. In this chapter we consider statistical normalization, which is the process of transforming the data points in a distribution in order to obtain a new distribution that is closer to a normal distribution. This is important because parametric test statistics can be used on (near-) normal distributions. Note however that the transformation must be sound, that is, the inherent relationships between the data points must not be destroyed by the transformation.

For further details on statistical concepts referred to in this chapter we refer the reader to Chapter 20.

6.1 Some illustrative examples

Consider a specific cod protein of interest in two situations: Cod in unpolluted water and cod in polluted water. The task is to examine if the protein differs in abundance between cod from the two situations. We therefore sample replicates (samples) from each of the two situations, and apply test statistics for comparing the population means.

Several such test statistics (see Chapter 20.5.2) require that the abundances from the two situations (populations) must be normally distributed (or nearly normally distributed), and some have the additional requirement that the variances must be equal for the two populations. These requirements are especially important when small samples are compared. It is therefore important to examine whether it is appropriate to assume that the samples are taken from normally distributed populations, and, if required by the test selected, to consider the assumption of equality of variances.

Computational and Statistical Methods for Protein Quantification by Mass Spectrometry,
First Edition. Ingvar Eidhammer, Harald Barsnes, Geir Egil Eide and Lennart Martens.
© 2013 John Wiley & Sons, Ltd. Published 2013 by John Wiley & Sons, Ltd.

In Chapter 5.6, we saw that a (relative) protein abundance can be calculated from multiple peptide occurrences. In the analysis it is often convenient to treat the different abundances from these peptides as normally distributed.

When searching for differentially abundant proteins, it is often appropriate to assume that a large majority of the proteins remain unchanged in abundance between situations, and that these unchanged proteins are normally distributed, often centered around zero on the logarithmic scale.

Note however that test statistics are available (referred to as *robust statistics*) that do not require normality and/or variance equality.

6.2   Non-normally distributed populations

Before describing tests for normality, we will examine the ways in which deviations from normality can occur.

6.2.1   Skewed distributions

The *skewness* of a distribution is a measure of its asymmetry, or its so-called *tailing*. One has two types (or directions) of skewness:

Positive skew   means that the distribution has a steeper decrease to the left than to the right. We say that the right *tail* is longer, and that the distribution is *right-skewed*. Figure 6.1(a) illustrates this.

Negative skew   means that the distribution has a steeper decrease to the right than to the left. The left tail is longer, the distribution is *left-skewed*. Figure 6.1(b) illustrates this.

6.2.2   Measures of skewness

Various formulas for measuring skewness have been proposed. An obvious constraint on the measures is that the skewness of a perfectly symmetrical distribution should yield zero for the metric.

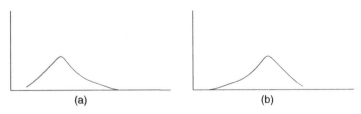

(a)   (b)

Figure 6.1   Example of distributions, positively skewed (a), and negatively skewed (b).

The first proposal was presented in Pearson (1895), and considers the difference between two location estimators for a distribution: The mean (μ) and the mode. Remember that the mean and mode should be equal for unimodal symmetrical distributions, yielding a difference of zero for that situation, thus satisfying the above mentioned constraint. The actual formula provides a relative measure of skewness by dividing the difference between mean and mode by the standard deviation (σ), yielding *Pearson's skewness coefficient*:

$$\frac{\mu - mode}{\sigma}.$$

Note that the direction of the skewness is given by the sign of the coefficient, with a value greater than zero indicating positive skew, while a negative value signals negative skew.

A simple method for estimating the mode of (unimodal) continuous data is to use *grouped frequency distribution*. In this approach a sliding window of a small size is used on the interval between the smallest and the largest value. The average of the smallest and the largest value in the interval for which the sliding window contains the highest number of values is selected as the mode.

Example Let the values be: 0.113, 0.152, 0.189, 0.223, 0.249, 0.262, 0.283, 0.314, 0.319, 0.342, 0.351, 0.372, 0.397, 0.412, 0.429, 0.458, 0.492, 0.516, 0.541, 0.572.

We set the window size to 0.05. Then the values 0.314, 0.319, 0.342, 0.351 are contained in a window. All other windows contain less than four values. The mode is therefore estimated to (0.314 + 0.351)/2 = 0.333.
△

References to other measures of skewness are found in the Bibliographic notes.

6.2.3   Steepness of the peak – kurtosis

The *kurtosis* of a distribution is a measure of its relative *steepness, peakness,* or *flatness*. The most frequently used definition for kurtosis calls for the normal distribution to have a kurtosis of zero, making the normal distribution, that is, with zero excess kurtosis *mesokurtic*. Distributions with thinner peaks (positive kurtosis; higher steepness) are called *leptokurtic*, while those with wider peaks (negative kurtosis; higher flatness) are called *platykurtic*. Figure 6.2 shows the various forms of kurtosis.

For leptokurtic distributions, most of the variance comes from a few extreme values, that is, a limited number of observations far away from the sharp peak around the mean, whereas for platykurtic distributions the variance is primarily derived from much more frequent but more modest deviations, that is, a larger number of values that lie in the flatter, broader bump around the mean.

Since the kurtosis (and the skewness) of a normal distribution is chosen to be zero, any substantial deviations from zero should raise concerns about the normality

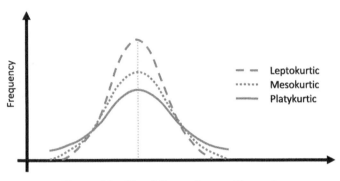

Figure 6.2   The different forms of kurtosis.

of the distribution. Tests for normality should then be used to ascertain whether the assumption of normality still holds.

6.3   Testing for normality

The most intuitive way of checking for normality is to consider a histogram of the data, and see if it appears to be symmetric. Intervals around the mean can then be explored, using the fact that intervals of length two, four, and six standard deviations cover probabilities 0.683, 0.954, and 0.997, respectively if normally distributed.

However, a long list of test statistics have been developed for verifying normality, see for example Farell and Roger-Stewart (2006). Many of these are implemented in standard statistical packages, returning a testable probability for the null hypothesis that the data are derived from a normal distribution.

For a brief introduction we here describe an often-used graphical tool based on quantiles, called a *normal probability plot*. Such a plot is a special form of the *quantile-quantile plot* (or *Q-Q plot*). (See Chapter 20.7 for a detailed description of quantiles.) A Q-Q plot can be used to check whether two samples come from the same (or similar) populations, or whether a sample comes from a specific distribution. It is this last application that we are interested in here, and the specific distribution is of course the normal distribution.

In a Q-Q plot the two axes represent quantiles of either the two samples to compare, or of the single sample and the chosen distribution to compare against. A point is plotted for the corresponding quantiles on each axis, and a correlation line can be drawn through the points. A correlation coefficient can also be calculated, indicating how good the correlation is. If the two samples come from the same distribution, or if the chosen ideal distribution provides a good model for the sample, the data points should form a straight line through origin, with a slope equal to one, as shown by the solid line in Figure 6.3. Alternatively, if the distributions are linearly related the points should approximate a straight line.

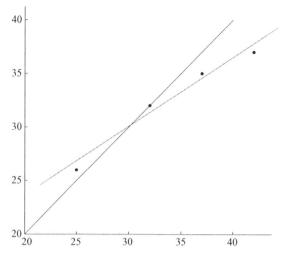

Figure 6.3   A Q-Q plot for the data in the example, with a correlation line drawn through the quintile points (the dashed line). We observe that the points are fairly well correlated by a straight line, but that it has some distance to origin. The distributions therefore seem to be linearly related in this case.

Example Let two samples be $\{20, 25, 25, 32, 35, 37, 42, 45, 58\}$ and $\{18, 19, 26, 28, 32, 32, 35, 37, 37, 43, 45\}$. We use $q = 5$, and calculate the resulting quintiles as $\{25, 32, 37, 42\}$ and $\{26, 32, 35, 37\}$ respectively. Figure 6.3 shows the corresponding Q-Q plot.

△

6.3.1   Normal probability plot

When a single sample is tested against a theoretical normal distribution, the Q-Q plot is called a *normal probability plot*. In this case, the calculated quantiles from the normal distribution are usually represented on the horizontal axis. If the theoretical and sample distributions are identical, the correlation line should intersect the vertical axis at zero, and the slope of the line should be one.

At this point two questions must be answered when investigating a sample for normality: Which normal distribution to test against, and when is normality satisfied?

Which normal distribution to test against?

We do not know the mean and standard deviation of the 'best' normal distribution to compare against, and are therefore not able to calculate the theoretical quantiles on the horizontal axis. In other words, the μ and σ for the theoretical distribution $N(\mu, \sigma)$ are unknown.

One approach could be to estimate these parameters by the sample mean and standard deviation, but it can be shown that using the standard normal distribution ($N(0, 1)$) is the best choice in the theoretical calculation for the horizontal axis. It is relatively easy to use the standard normal distribution however, since a quantile q_N of $N(\mu, \sigma)$ is related to the corresponding quantile q_S of the standard normal distribution through $q_N = \mu + \sigma q_S$ (remember that if a variable Y has a normal distribution $N(\mu, \sigma)$, the variable $Z = (Y - \mu)/\sigma$, describes the standard normal distribution).

If the sample does originate from a normal distribution $N(\mu, \sigma)$, the points on the normal probability plot against the standard normal distribution will form a straight line. Also, from the equations above we see that the intersection of the line with the vertical axis (at $q_S = 0$) occurs at value μ, and that the slope of the line is σ. It is therefore possible to estimate (μ, σ) from the correlation line of the plot in such cases.

Calculating the standard normal quantiles   When we have n values we use $n + 1$-quantiles. The i-th $n + 1$-quantile q_i is then calculated from the standard normal distribution as

$$\frac{i}{n + 1} = P(X \le q_i).$$

The 4-th 9 quantile of the standard normal distribution is for example

$$\frac{4}{9} = 0.4444 = P(X \le -0.14).$$

Example   Let the 9-quantiles of a set of observations be 12.1, 12.7, 13.5, 14.4, 14.7, 15.2, 16.4, 17.2. We want to test how good this can fit a normal distribution. The theoretical quantiles from $N(0,1)$ are calculated as: -1.22, -0.77, -0.43, -0.14, 0.14, 0.43, 0.77, 1.22.

Figure 6.4 shows a normal probability plot for the example. If we accept that the correlation is good enough, we can estimate the mean of the population to be 14.62 (the intersection with the vertical axis), and the standard deviation to be 1.69 (the slope of the correlation line). For comparison, we can calculate the sample mean in the traditional way to 14.53, and the sample standard deviation to 1.75.
△

When is normality satisfied?

We can calculate a correlation coefficient for the probability plot, and accept normality if the correlation is strong enough.

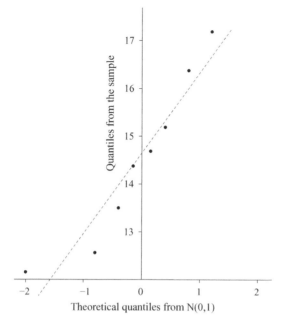

Figure 6.4   A normal probability plot.

6.3.2   Some test statistics for normality testing

There are just a few test statistics for testing normality of a set of data. We will here first describe the Kolmogorov–Smirnov test, mainly due to its simplicity.

Two cumulative distributions C and C' of the ordered data $X_{(1)}, X_{(2)}, \ldots, X_{(n)}$ are calculated and compared to the cumulative distribution F of the considered (normal) distribution.

- $C_i = \frac{i}{n}, 1 \leq i \leq n$;

- $C'_i = \frac{i-1}{n}$.

The test statistic is then defined as:

$$D = \max_{1 \leq i \leq n} (\max (F(X_{(i)}) - C'_i, C_i - F(X_{(i)}))).$$

To test for normality we formulate the null hypothesis that the data is sampled from a normal distributed population. The calculated D-value is compared to a critical value, and if it is larger the null hypothesis is rejected, otherwise it is accepted. How to calculate the critical values is outside the scope of this book.

The Kolmogorov–Smirnov test requires that the parameters of the normal distribution to test against are known, they cannot be estimated from the data. It is also more sensitive at the centre of the distribution than at the tails. Therefore other test statistics are usually preferred.

One option is the Shapiro–Wilk W-statistic calculated as follows:

$$W = \frac{(\sum_{i=1}^{n} w_i X_{(i)})^2}{\sum_{i=1}^{n} (X_i - \bar{X})^2}.$$

How to calculate the weights w_is can be found in Royston (1982). Note that this test does not work well for data containing multiple identical values.

Yet another test is the D'Agostino–Pearson omnibus test. It first calculates the skewness and kurtosis of the data, and then calculates a value based on how far each of these is from values expected if the data is normally distributed.

6.4 Outliers

Before statistical analysis is performed on observed data, the data should be investigated for the presence of *outliers*. An outlier in a set of observed data is a value that seems inconsistent with the rest of the data. There are several sources for outlier observations:

- they can come from recording or measuring errors;

- they can come from contaminations in the sample;

- they can show that the expected distribution is not a good match to the actual data.

In the context of protein quantification however, it is important to have in mind that *significantly differentially abundant proteins will occur as outliers*. The main issue is that the distinction between an outlier and a differentially abundant protein is essentially impossible to make *a priori*, and possibly impossible to make *a posteriori* as well without follow-up experiments or external (orthogonal) information.

How to cope with the outliers therefore depends on the context, below are two examples:

Calculating the mean   When assuming that the data comes from a specific distribution (often normal), identified outliers can be removed before calculation. An example of this is calculating the abundance mean from a set of peptides assumed to originate from the same protein. All the peptide values should ideally be the same, and the reasons for an outlier can be that the peptide does not come from the considered protein, or that it comes from several proteins.

Investigating the underlaying distribution   The dataset without the outliers can be investigated to see if the (underlaying) distribution is (for example) normal, and test statistics assuming normal distribution can be used. An example of this is testing for differentially abundant proteins.

In general it is problematic both to identify outliers and to determine their sources, and to figure out how to handle them. One rule is that if the source of an outlier is not determined one should not remove it, but handle it. Outliers may result in problems with statistical inference based on the normal distribution assumption and one may resort to using nonparametric statistics instead. If the outliers are not due to measurement errors one can perform sensitivity analysis in the sense of seeing if the results (and conclusions) change when the outlier(s) are included/excluded. In any case, outliers should be reported and discussed.

Most of the test statistics developed for identifying outliers assume that the overall distribution is a normal distribution. Before these statistics can be used, the observed data should therefore be shown to be fairly well approximated by a normal distribution.

A test statistic for outlier identification must formulate an underlying definition of how to characterize an outlier. The statistics described below use formulas related to mean, median, or quantiles to achieve this definition.

Example For some of the statistics we describe we provide an example. These all use the dataset 4, 6, 8, 8, 10, 12, 13, 24. Though this is a small dataset it should demonstrate how the different statistics work.

The sample mean and standard deviation of this set are 10.63 and 6.16, respectively.

$\triangle$

6.4.1   Test statistics for the identification of a single outlier

Numerous test statistics exist to test for a single outlier, of which we will mention some of the most commonly used. The null hypothesis in these cases is typically given as: There are no outliers.

Grubb's test

A normal distribution of the data is assumed. The test statistic is defined as[1]

$$G = \frac{|X_{ext} - \bar{X}|}{S},$$

where

- X_{ext} is the value in the sample furthest away from the sample mean;

- $\bar{X}$ is the mean of the sample;

- S is the standard deviation of the sample.

A T distribution is used for calculating *critical values*.

---

[1] Note that this is another statistic than the G-statistic in Chapter 8.2.2.

If the significance level for rejection is α, the critical value for a two-sided test is

$$c(\alpha) = \frac{n-1}{\sqrt{n}} \sqrt{\frac{t^2_{\alpha/(2n)}}{n-2+t^2_{\alpha/(2n)}}},$$

where the T distribution with $df = n - 2$ is used.

Example Choose $\alpha = 0.05$. Then $t_{0.05/16} \approx 3.8$ for $df = 6$. Thus $c(0.05) \approx 2.1$. This is less than the value of G for 24 (2.17), hence 24 is considered an outlier. Then use $\alpha = 0.01$. $t_{0.001/16} \approx 6$. Thus $c(0.01) \approx 2.3$. This is higher than the value of G, so 24 is not considered an outlier at this confidence interval.
$\triangle$

If only the minimum value, or the maximum value, is tested for being an outlier (one-sided test), the critical value is

$$\frac{n-1}{\sqrt{n}} \sqrt{\frac{t^2_{\alpha/n}}{n-2+t^2_{\alpha/n}}},$$

where $df = n - 2$.

Grubb's test is also called the *extreme studentized deviate, ESD*. Tables with critical values for t_α can be found on the web.

Dixon's test

Again, a normal distribution is assumed for the dataset. Dixon's test comprises a set of test formulas. It tests either $X_{(1)}$ (the minimum value) or $X_{(n)}$ (the maximum value) for being an outlier, taking into account that there might be additional outliers. Ratios

$$r_{ij} = \frac{X_{(1+i)} - X_{(1)}}{X_{(n-j)} - X_{(1)}} \text{ or } = \frac{X_{(n)} - X_{(n-i)}}{X_{(n)} - X_{(1+j)}}; 1 \leq i \leq 2; 0 \leq j \leq 2$$

can be calculated for testing $X_{(1)}$ or $X_{(n)}$, respectively.

- i is the number of possible outliers at the same side as the value being tested;
- j is the number of possible outliers at the other end.

Example $r_{10} = \frac{X_{(2)} - X_{(1)}}{X_{(n)} - X_{(1)}}$ tests $X_{(1)}$, assuming no other outliers; $r_{21} = \frac{X_{(n)} - X_{(n-2)}}{X_{(n)} - X_{(2)}}$ tests $X_{(n)}$, assuming $X_{(n-1)}$ and $X_{(1)}$ are also outliers.
$\triangle$

Table 6.1   Critical values for r_{10} for a two-sided Dixon's outlier test, from Rorabacher (1991).

n	90%CL	95%CL	99%CL
3	0.941	0.970	0.994
4	0.765	0.829	0.926
5	0.642	0.710	0.821
6	0.560	0.625	0.740
7	0.507	0.568	0.680
8	0.468	0.526	0.634
9	0.437	0.493	0.598
10	0.412	0.466	0.568

r_{ij} is then compared to critical values. Different critical values are calculated for different combinations of i and j, and different confidence intervals. If r_{ij} is greater than the critical value used, the tested value is considered an outlier.

Critical values for r_{10} are shown in Table 6.1 for different confidence intervals when a two-sided test is used. The table is copied from Rorabacher (1991), where critical values for the other ratios and other confidence intervals can be found for sample sizes up to 30.

Example For our example we find $r_{10} = \frac{24-13}{24-4} = 0.55$. If we use $\alpha = 0.05$ we see that the critical value is 0.526, hence 24 is considered an outlier.
△

Larger samples are required for larger i and j. Dixon formulated the following rules:

- for $3 \leq n \leq 7$, r_{10} should be used;
- for $8 \leq n \leq 10$, r_{11} should be used;
- for $11 \leq n \leq 13$, r_{21} should be used;
- for $14 \leq n$, r_{22} should be used.

Using r_{10} is often referred to as Dixon's Q-test ($Q = r_{10}$). It is a simple and often used outlier test for small datasets. It is also appropriate for larger datasets when only one outlier is expected.

Use of quartiles

This is a nonparametric statistic that does not require normality. The quartiles (see Chapter 20.7) Q_1 and Q_3 can be used to estimate the spread of the data, akin to the use of the parametric standard deviation statistic for normally distributed data. Any value at a distance larger than or equal to $k(Q_3 - Q_1)$ from this interval is then considered an outlier, with k a chosen constant. More formally, X is considered an outlier if

$$\frac{X - Q_3}{Q_3 - Q_1} > k \text{ or } \frac{Q_1 - X}{Q_3 - Q_1} > k.$$

This means that X is an outlier if it is outside the interval

$$[Q_1 - k(Q_3 - Q_1), Q_3 + k(Q_3 - Q_1)]$$

Note that the distance $(Q_3 - Q_1)$ is often written as the *interquartile range, (IQR)* or the *interquartile distance, (IQD)* and that k is typically chosen to be 1.5.

Example For our example data we find $Q_1 = 7$, $Q_2 = 9$, $Q_3 = 12.5$. We choose $k = 1.5$ and calculate $\frac{24-12.5}{12.5-7} = 2.1$, hence 24 is considered an outlier.
$\triangle$

6.4.2   Testing for more than one outlier

Testing for more than a single outlier can be done in two different ways:

1. Iteratively apply a test for a single outlier, taking care to remove values considered to be outliers from the dataset before performing the test again on the reduced dataset in the next iteration.

2. Use statistics that are specifically developed for several outliers.

Note that Dixon's test described above can be used for up to four outliers $(i = j = 2)$.

Rosner's test for multiple outliers

Rosner's test is a generalization of Grubb's test, valid for up to ten suspected outliers, and for samples with 25 or more observations.

First an upper limit $k \leq 10$ is defined for the number of potential outliers. An iterative procedure is then used that removes an additional potential outlier from the sample in each iteration. Let

- $\bar{X}^{(i)}$, $S^{(i)}$ be the mean and standard deviation of the sample (of size $n - i + 1$) when $i - 1$ values are removed, $i = 1, \ldots, k$;

- $X^{(i)}$ be the value in the remaining sample (of size $n - i + 1$) lying farthest away from $\bar{X}^{(i)}$.

Then define the test statistics

$$R^{(i)} = \frac{|X^{(i)} - \bar{X}^{(i)}|}{S^{(i)}}, \; i = 1, \ldots, k.$$

Each $R^{(i)}$ can be tested against a critical value $R^{(i)}_{n,\alpha}$, where α is the significant level. If $R^{(i)} > R^{(i)}_{n,\alpha}$ then the i first removed values are all considered outliers, assuming a normal distribution. The number of outliers j is determined as

$$j = max_i\{0 < i \leq k : R^{(i)} > R^{(i)}_{n,\alpha}\}.$$

Values for $R^{(i)}_{n,\alpha}$ can be found in Rosner (1983).

Use of box plots

Box plots are used as a visual tool for identifying outliers. An example is shown in Figure 6.5.

Box plots are further described in Chapter 20.7.3. For the box plots constructed by SPSS for showing outliers the T-bars are called *fences* or *whiskers*. The distances to the fences from the box are 1.5 times the height of the box, or to the minimum or maximum value in this range. For normally distributed data, approximately 95 % of them are expected to lie between the fences. The points over and below the fences are data values considered as outliers. If a value is more than three times the height of the box away from the box, it is considered as an extreme outlier, and marked as a star.

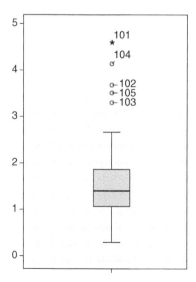

Figure 6.5   A box plot of a sample of 105 numbers of which there are five outliers. Constructed by SPSS.

6.4.3   Robust statistics for mean and standard deviation

Robust statistics are statistics that are largely unaffected by outliers, thus avoiding the necessity to remove outliers. One should however always consider the efficiency of these statistics as estimators before deciding between outlier removal and the use of more efficient parametric statistics on the one hand, and the use of less efficient robust statistics on the other hand. We here consider robust statistics for mean and standard deviation.

Robust statistics assume that the underlying distribution is nearly normal with some outliers, which correlates well with the expected results from a typical quantitative proteomics experiment. As always, care must therefore be taken because the application of robust methods on data that break this assumption can lead to wrong results.

The median

A simple way of estimating the location of a distribution is by using the median. The median can then in turn be extended to estimate the scale of the distribution. Let d_i be the absolute value of the difference between X_i and the median. Then sort the d_is in ascending order, and find the median of these values. This is denoted the *Median Absolute Deviation (MAD)*. The MAD can be used as an estimator of the scale of the distribution.

Example Suppose that we have the dataset 5, 14, 16, 21, 44. The mean and standard deviation of all observations are 20 and 14.6. When 44 is removed as an outlier, these statistics become 14 and 6.7. The median is 16, and the MAD is 8. Hence 16 is a robust estimated value for 14, and 8 for 6.7.

$\triangle$

Trimmed means

Trimmed means are calculated by ignoring a certain percentage of the highest and lowest values before calculating the mean. A $n\%$ trimmed mean means that $\frac{n}{2}\%$ of the highest and lowest values are removed prior to calculating the mean. Thus, this is equivalent to removing a specific set of outliers from the sample.

Huber's method

A more advanced method is *Huber's* method, Roy (2001), which uses more of the information contained in the values than is achieved by simple trimming. In an iterative procedure it transforms the extreme values in the dataset through a form of transformation called *Winsorization*. Possible outliers are first identified as all values outside the symmetrical interval around a location estimator m_0 (most commonly the mean or median) with a width of $2k$ the estimated scale s_0 (typically the standard deviation or MAD). The parameter k is usually chosen to be 1.5. Any measurement in the dataset that falls outside the interval thus defined,

$$[m_0 - ks_0, m_0 + ks_0]$$

is considered an outlier, and its value is replaced by the closest interval edge ($m_0 - ks_0$ or $m_0 + ks_0$). In the next iteration, the revised values in the dataset are used to recalculate a location and scale estimator, m_1 and s_1. These are then compared to the previous estimates and if their difference is smaller than a preset convergence criterion ϵ, the iterations end. Otherwise the calculation continues as before, with all outliers outside the interval

$$[m_1 - ks_1, m_1 + ks_1]$$

replaced by the values $m_1 - ks_1$ or $m_1 + ks_1$. When convergence is reached, the current m_i and s_i can be used as robust location and scale estimators. The convergence criterion ϵ is typically set at a small value such as 10^{-6}.

Note that the Winsorization process essentially squeezes the data into an ever shrinking interval, while the step size (the size of the shrinking) is reduced at every step. Contrary to the trimming approach, the outliers are maintained in this method but are continuously reassigned new values based on the largest acceptable values.

Identifying outliers

We mentioned that robust statistics more or less ignore outliers. If we still want to identify the outliers however, they will most often be detectable by calculating the z-score of the data, using the computed estimators for mean and standard deviation.

6.4.4   Outliers in regression

Regression analysis, see Chapter 20.9 is highly sensitive to outliers, often resulting in erroneous or false conclusions. The simplest way for outlier detection in regression analysis is a *residual plot*, where the independent variable is plotted on the horizontal axis, and the residuals on the vertical axis. For a good correlation the points should lie around the line $y = 0$.

Cook's distance

One approach for considering outliers in regression is to take into account the influence each value has on the parameters of the regression line, for example, using *Cook's distance*, Cook (1977).

6.5   Variance inequality

It has been determined empirically that variance inequality is an issue in proteomics, since low abundant proteins generally have larger variance than high abundant proteins, as illustrated in Figure 5.8, p. 72. This influences which test statistics can be used. Furthermore, in correlation analyses this can result in too much emphasis being put on high abundant proteins.

This phenomenon is often called heteroskedasticity, or heterogeneity of variance. The complementary notation is homogeneity of variance.

6.6   Normalization and logarithmic transformation

Normalization in this context means transforming a non-normal set of data (observations) by a mathematical function such that it (or the population from which they are sampled) achieves a normal distribution. The data can then be analyzed by statistics suited for normal distributions (parametric statistics).

Some of the most commonly used transformations will also perform *variance stabilization*, thus achieving equal variance across the dataset (homogeneity).

The most often employed transformation is the *power transformation*, typically a square root transformation, logarithmic transformation, or inverse transformation.

Square root   transformation is used to normalize negative skewed (right-skewed) datasets. Such transformations will for example convert data with a Poisson distribution to a normal distribution. Note that this transformation cannot be used for datasets with negative values.

Square   transformation is used to normalize positive skewed (left-skewed) datasets. Squaring should not be used for datasets with negative values.

Logarithmic   transformation is used to normalize data with an exponential distribution. Any distribution that will become normal by a logarithmic transformation is said to be *log-normal distributed*. This transformation cannot be used for datasets containing negative values or zeros.

Inverse   (or 1/X) transformation can also be used to normalize data with an exponential distribution.

We will here describe the logarithmic transformation in more detail, especially since this transformation also serves other purposes.

6.6.1   The logarithmic function

The standard notation for the logarithm, $\log_b X$, with base b and real value X is well defined for $b, X > 0$, as

$$X = b^{\log_b X}.$$

It is a function of the two variables b, X. When the base b is fixed, for example, $b = 2, b = 10, b = e$, one *definite logarithm* is obtained, which is a function of one variable, X. It has the following properties:

- it is defined for all $X > 0$;
- $log_b(1) = 0$;
- $log_b(XY) = log_b(X) + log_b(Y)$;
- $log_b(X/Y) = log_b(X) - log_b(Y)$.

The logarithms of a number X of two different bases b, c differ by a constant factor only, $log_b c$. This is shown by the equation

$$\log_b X = \log_c X \log_b c = k \log_c X, \tag{6.1}$$

for a constant k. This equation follows from

$$X = b^{\log_b X} = c^{\log_c X} \Rightarrow \log_b X \log_b b = \log_c X \log_b c \Rightarrow \log_b X = \log_c X \log_b c.$$

When the base does not matter, we can use the *indefinite logarithm*. The notation for the indefinite logarithm is $Log(X)$ or $[\log X]$, but the simple base-less $\log X$ is also used in more informal presentations. However, often when $\log X$ is used, an implicit base is assumed that should be clear from the context. Going deeper into indefinite logarithms is outside the scope of this book.

6.6.2   Choosing the base

The choice of the base to use depends on the data and the analysis that is planned, and is also a matter of convenience. From Equation 6.1 we see that it is only the values that are different, so the choice of the base does not effect the outcome of the analysis. The mathematical properties of the natural logarithm (ln) and of e (such that the derivative of $ln(X)$ is $\frac{1}{X}$ and the derivative of e^X is e^X) make this a convenient base to use in many applications. However, since the abundance changes of proteins (and mRNAs) are often expressed as fold changes, base two is most commonly used.

6.6.3   Logarithmic normalization of peptide/protein ratios

Quantitative proteomics experiments typically result in the calculation of ratios between two situations for a lot of peptides or proteins. It has been shown that for non-differentially abundant proteins these are log-normally distributed when the number of peptides or proteins is large. A logarithmic transformation will thus turn these ratios into a normal distribution.

6.6.4   Pitfalls of logarithmic transformations

Logarithmic transformation might in some cases produce outliers that were not outliers in the original data, and then bias the correlation.

Example Let the original data be 0.02, 0.6, 0.9, 1.3, 1.8, 2.3, 2.6, 3.0. 0.02 is not an outlier here using Dixon's Q-test. If we take the (natural) logarithm of these values however, -3.91 (ln 0.02) will be considered an outlier (the difference between -3.91 and its nearest value is extreme).
$\triangle$

Outliers may be produced when there are (measuring) errors in the data, and these errors are additive. Having *additive errors* means that the error is simply added to the true value, and does not depend on the true value. *Multiplicative errors* on the other hand, means that the error depends proportionally on the true value. For multiplicative errors the logarithmic transformation will make the errors symmetric, and unexpected outliers as illustrated above will not occur. This is shown below.

Let y be a correct value, and δy an additive error such that the measured value is $y_m = y + \delta y$. Performing logarithmic transformation will produce $Log(y + \delta y)$ instead of $Log(y)$. The difference between these two values decreases with increasing y.

Example Let $\delta y = 0.2$ and Log is ln. Then for $y = 1$, we get ln 1.2 − ln 1 = 0.18, while for $y = 10$ this gives ln 10.2 − ln 10 = 0.02
$\triangle$

Consider then a multiplicative error of factor k. The difference between the measured value and the correct value is $Log(y + ky) - Log(y) = Log(y(1 + k)) - Log(y) = Log(y) + Log(1 + k) - Log(y) = Log(1 + k)$, showing that the error in the transformed values is independent of the values themselves.

From this we conclude that logarithmic transformation can be dangerous when the measurements of the values contain significant additive errors, but is appropriate for multiplicative errors.

6.6.5   Variance stabilization by logarithmic transformation

Transformation of the measured abundances is performed here in order to obtain a symmetric distribution with constant variance over the range of measured abundances. If the increase in variance with increasing abundance is small, a logarithmic transformation has such a variance stabilization effect. If the increase is large a transformation with the n-th root may achieve homogeneity. These types of transformations can be combined in a more general function, Karp et al. (2010).

Let

- y be a correct abundance value;

- z be the measured or determined value.

Then a more generalized log transformation is

$$t_c(z) = \log(z + \sqrt{z + c}),$$

where c is a constant to be estimated from the data, that can be calculated using software. This transformation is:

- similar to the logarithmic transformation at high abundances, where the multiplicative errors dominate;
- similar to a linear transformation at low abundances, where the additive errors dominate;
- interpolating in between these errors at middle abundances.

In relative quantification it is often the ratio between two abundances that is transformed, $\log \frac{z_1}{z_2}$. Then this generalized log ratio is defined as

$$\log \frac{z_1 + \sqrt{z_1^2 + c^2}}{z_2 + \sqrt{z_2^2 + c^2}}.$$

6.6.6   Logarithmic scale for presentation

Proteomics data often covers a large range of values, possibly spanning several orders of magnitude. In order to clearly represent all these data in one display, the logarithmic scale is often appropriate.

Example Let the following be values from a collection of data:

$$1.0,\ 1.2,\ 3.6,\ 6.2,\ 12.5,\ 19.7,\ 40.3,\ 70.0,\ 85.6,\ 98.2$$

If plotted in a one-dimensional linear plot on one page it would be difficult to distinguish between the first two numbers. A logarithmic transformation using base 10 on the other hand results in:

$$0,\ 0.1,\ 0.6,\ 0.8,\ 1.1,\ 1.3,\ 1.6,\ 1.8,\ 1.9,\ 2.0,$$

where all values will be clearly visually distinguishable.
$\triangle$

Interestingly, new research in cognitive psychology indicates that presenting numbers in a logarithmic scale, that is, ratio based, is the most intuitive way, inherited through evolution. Through education however, we have become used to linear presentation, see for example Dehaene et al. (2008). That logarithmic presentation is the most natural way can also be argued by using *Weber's law*, which says that the just observable difference is a constant proportion of the original stimulus value. For example, this means that if a person hears a sound of 10 units, and observes a change when the sound is increased to 11 units, then an increase from 100 units must be 10 units in magnitude before the change is observed.

When we have two variables, and we want to illustrate or analyze relations between them, we can make use of three different, commonly used plot types:

Linear plot   using the original values.

Logarithmic plot   using values that are logarithmically transformed along both axes before plotting.

Semi-logarithmic plot   using logarithmically transformed values along only one of the axes before plotting.

If the original data (X, Y) satisfy a relation $Y = a^{bX}$, a logarithmic transformation of Y results in $Log(Y) = bXLog(a) = kX$, thus showing a linear relation on a semi-logarithmic plot.

If the original data satisfy a relation $Y = X^a$, logarithmic transformation of both X as well as Y results in $Log(Y) = aLog(X)$, thus showing a linear relation on a logarithmic plot.

Furthermore, if the significance of differences between values (or changes in values) depends on the value and not on the value of the difference, a logarithmic scale is appropriate. For example this means that a change from one to two has the same significance as a change from 100 to 200, consistent with Weber's law.

The acidity of a watery solution provides an example of a logarithmic scale, where the acidity (pH) is measured as the negative ten base logarithm of the concentration of H^+ ions in the solution. Another example is the Richter magnitude scale for energy contained in an earthquake, that uses a ten base logarithm of the waves measured by a seismograph.

6.7   Exercises

1. Assume that we have performed an experiment where the relative abundances of a set of proteins was as follows:

   0.76, 0.84, 0.85, 0.88, 0.93, 0.96, 0.98, 0.99, 1.01, 1.02, 1.04, 1.06, 1.08, 1.12, 1.15, 1.18, 1.24, 1.35, 1.92, 2.12.

   Often experiments are performed such that the relative abundances of proteins that are not differentially abundant are log-normal distributed. For later statistical analysis we therefore calculate $\log_2$ of the ratios:

   $-0.40, -0.25, -0.23, -0.18, -0.10, -0.06, -0.03, -0.01, 0.01, 0.03, 0.06, 0.08,$ 0.11, 0.16, 0.20, 0.24, 0.31, 0.43, 0.94, 1.08.

(a) Calculate mean, standard deviation, median, and mode for the logarithmic values. Use grouped frequency distribution for calculating the mode, with subintervals of 0.05.

(b) Calculate the skewness using $(\bar{X} - median)/S$. Is there a skewness in the data?

(c) Considering the values it seems that there are two outliers to the right. Remove these, and recalculate mean, standard deviation, median and mode, and the skewness. Did the removal of the outliers have any effect?

2. The 7-quantiles of a set of observations are 20.1, 22.2, 23.7, 24.5, 26.9, 28.1.
   (a) Calculate the corresponding quantiles from the standard normal distribution.

   (b) Plot the quantiles in a normal probability plot. Draw a correlation line and estimate the mean and the standard deviation.

3. The following values are observed: 4, 7, 9, 12, 14, 20. Use Grubb's and Dixon's tests to test if 20 is an outlier. Use $\alpha = 0.05$.

4. Assume that we have observed 26 values 1.28, 3.12, 4.12, 4.45, ... , 11.18, 13.51. We suspect that there can be up to three outliers, and will use Rosner's test. We will use a significant threshold of $\alpha = 0.05$, and the critical values $R^{(i)}_{26,0.05}$, $i = 1, 2, 3$ are 2.84, 2.82, 2.80 respectively. We perform the following calculations:

- $\bar{X}^{(1)} = 7.65$, $S^{(1)} = 2.16$;

- $\bar{X}^{(2)} = 7.95$, $S^{(2)} = 1.92$;

- $\bar{X}^{(3)} = 7.72$, $S^{(3)} = 1.79$.

  Determine the outliers.

6.8   Bibliographic notes

Non-normal distributions   Pearson (1895), von Hippel (2005).

Test for normality   Royston (1982) (Shapiro-Wilk), Farell and Roger-Stewart (2006), Chiogna et al. (2009).

Outliers Grubb's test   Grubbs (1969), Verma and Quiroz-Ruiz (2006a).

Outliers Dixon's test   Dixon (1953), Rorabacher (1991), Verma and Quiroz-Ruiz (2006b).

Outliers Rosner's test   Rosner (1983).

Variance stabilization   Huber et al. (2002), Durbin et al. (2002), Karp et al. (2010), Box and Cox (1964).

Power transformations   Box and Cox (1964).

7

Experimental normalization

Quantitative proteomics experiments involve the comparison of protein abundances from several quantification runs. If we can ensure that the sampling process for each run is identically executed, and that the required instruments work in exactly the same way for every sample, we could compare the resulting profiles directly. However, in reality there will always be variations in sample handling and instrument operation, resulting in artefactual variations in the profiles. Some form of normalization should therefore be performed across the experiments.

7.1 Sources of variation and level of normalization

Differences in the resulting profiles can therefore come from different sources:

1. Real differences in protein abundances for the situations we are comparing.

2. Real differences in protein abundances due to external factors not considered in the experiment, for example, changes in temperature or the digestion of different food items for the sample subjects.

3. Differences due to random noise or contaminants in the preparation or measurement process, for example, random variations in the measuring instrument or contamination that affects one (or more) of the samples but not the others.

4. Differences due to systematic shifts or drifts in the preparation or measurement process, for example, a measurement in one of the situations systematically measures a 10 % higher abundance than in the other situations.

Differences due to the second point have to be controlled during the processing of the experimental data, that is, by trying to keep the research subject in the same state or condition for all the situations, while differences due to the third point should be

Computational and Statistical Methods for Protein Quantification by Mass Spectrometry,
First Edition. Ingvar Eidhammer, Harald Barsnes, Geir Egil Eide and Lennart Martens.
© 2013 John Wiley & Sons, Ltd. Published 2013 by John Wiley & Sons, Ltd.

taken care of in the statistical analysis. Finally, the handling of differences due to the last point is the goal of experimental normalization and this will now be discussed in more detail.

It is important to realize that the actual systematic shift is unknown. Based on experience and common assumptions, we can however select the type of normalization that seems the most appropriate.

The type of normalization required has strong similarities to the normalization needed for gene expression microarray data, for which a number of different normalization methods have been developed. Knowledge from the field of microarrays is therefore employed when considering normalization methods for proteomics mass spectrometry data.

In the context of proteomics we make certain assumptions in order to perform normalization:

1. The samples to be compared are comparable, meaning that they contain essentially the same proteins.

2. Only a small subset of the proteins have different abundances in the samples, that is, most proteins have a 1:1 ratio.

3. There are enough proteins with unchanged abundances such that the differently abundant proteins can be considered outliers as compared to the overall distribution of protein abundances.

4. There is an overall log-normal distribution of the ratios.

Figure 7.1, illustrates a typical quantification experiment, and can be used as a reference to discuss the occurrence of unwanted variation and corresponding normalization procedures.

As an example assume that two sets of biological replicates for two situations (A and B) are given. There may be variations in:

- the biological replicates;

- the sample extraction, such that the samples may contain more variations than resulting from the biological variations;

- the sample preparation;

- the sample combination;

- the sample digestion;

- the sample labeling (the exact occurrence of the labeling step depends on the approach);

- the LC-MS/MS analysis

  - the chromatography;

  - the MS spectrum acquisition;

  - the MS/MS spectrum acquisition.

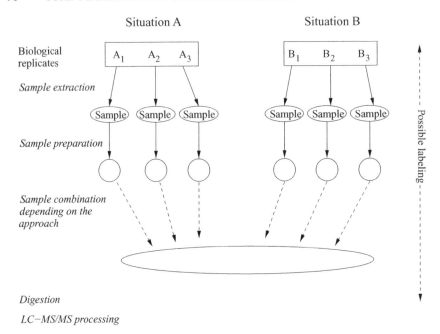

Figure 7.1   A model of a typical quantification experiment.

In most methods, protein abundances are calculated by combining several measured or calculated intensities or abundances. Normalization can thus be performed at different levels. The lowest level concerns the normalization of spectra, followed by the intermediate level of normalization of peptides, with the normalization of protein abundances as the highest level.

7.2   Spectral normalization

A fundamental operation in protein quantification is to compare the intensities of corresponding peaks, for example, peaks from the same peptide (MS spectra) or peaks from the same fragment (MS/MS spectra) in two or more spectra. Various undesirable effects during the creation of the spectra can influence the observed raw intensities. These effects should be eliminated before the comparison, and this is the aim of the normalization.

The various methods used for spectral normalization can be classified as *collective* or *individual*, Norris et al. (2007). In the collective approach each spectrum is normalized against a reference spectrum constructed from all the spectra. While in individual normalization each spectrum is normalized independently. Individual normalization is the dominant approach for spectral normalization, and will be the focus in the further description.

Note that though it is generally reasonable to compare spectra from the same molecule (protein or peptide), the number of peaks in the spectra to compare can vary slightly.

7.2.1   Scale based normalization

Scale based normalization means that the raw intensities are divided by a value calculated separately for each spectrum by a specific function. Therefore this is also termed relative normalization. Different functions results in different scale based normalization methods.

Let T be a spectrum, and t be the intensity of a peak in T. Furthermore, let $N(t)$ be the normalized intensity of t. Then

$$N(t) = \frac{t}{F(T)},$$

where $F(T)$ is derived from T. $F()$ is then what is expected to be invariant across the spectra.

This builds on the assumption that the measured intensities increase at the same ratio as the real abundances of the molecules (peptides). Let a_1, a_2 be the abundances of two peptides, and e_1, e_2 be the measured intensities of two corresponding peaks, the assumption is then that $a_1/a_2 = e_1/e_2$.

As explained in Chapter 5.8 it is most often relative intensities that are used for the quantification. Suppose that we have two spectra T_1 and T_2, and have found that a peak with intensity t_1 in T_1 corresponds to a peak with intensity t_2 in T_2, for example, representing the same peptide. Then the relative intensity is determined as

$$\frac{N(t_1)}{N(t_2)} = \frac{t_1/F(T_1)}{t_2/F(T_2)} = \frac{t_1}{t_2}\frac{F(T_2)}{F(T_1)} = \frac{t_1}{t_2}C.$$

Here C is a value that determines which relative intensities become substantially different from 1.

Several functions have been proposed for F, depending on what is assumed to be most useful as a normalizer. Note that the different functions result in different values of C, which may have consequences for the further analyses.

Example   We have two spectra with two pairs of corresponding peaks (from two peptides). Let the raw intensities of the peaks in the first spectrum be $t_{11} = 24$, $t_{12} = 15$, and in the second $t_{21} = 18$, $t_{22} = 18$. Let the threshold for the intensities being substantially different be a fold change of 2.

First consider a normalization where C becomes 2.1. Then $\frac{N(t_{11})}{N(t_{21})} = 2.8$, hence the difference between the normalized values becomes substantial. $\frac{N(t_{12})}{N(t_{22})} = 1.75$, hence not a substantial difference.

Then consider a normalization where C becomes 0.45. Then $\frac{N(t_{11})}{N(t_{21})} = 0.6$ (not a substantial difference), and $\frac{N(t_{12})}{N(t_{22})} = 0.37$ (substantial).

Though such large difference between the value of C for two different forms of F will be extremely rare it illustrates the phenomenon.

$\triangle$

We now consider a spectrum T with n peaks where the intensities are $t_1, t_2, \ldots, t_n$, and use this to explain the most used scale based normalization methods.

Normalization using total spectral intensity

This normalization builds on the assumption that the total intensities are similar for all spectra. The formula is

$$F(T) = \sum_{i=1}^{n} t_i$$

and is termed the TIC (Total Ion Current) of the given spectrum.

Note that this is a special case of the *p-norm*

$$F(T) = \left(\sum_{i=1}^{n} t_i^p \right)^{\frac{1}{p}}$$

for $p = 1$.

Normalization using the vector norm

The formula for this is

$$F(T) = \sqrt{\sum_{i=1}^{n} t_i^2}$$

that is, the p-norm for $p = 2$.

For this normalization the highest intensity peaks have more impact on the normalization than using TIC. This means that for spectra with a couple of large peaks the vector norm will give smaller normalized intensities than for spectra with more similar peak intensities, when the TICs are similar. How this effects the relative intensities (increase or decrease) as compared to using TIC, depends on the peak intensities, see Exercise 2.

Normalization by highest intensity

The invariant assumption is that the abundance of the most abundant molecule in the samples where the spectra are taken from is the same. For this the formula is

$$F(T) = \max_{i \in [1,n]} t_i.$$

This is the p-norm for $p \to \infty$, and demonstrates the increasing impact of higher intensity peaks for increasing p.

Dividing the spectra into bins

When $F()$ is not invariant over the whole spectrum it may still be invariant at distinct subspectra. The m/z-axis is then first divided into bins, thus introducing subspectra, B. Then to normalize by highest intensity $F(T) = \max_{j \in B} t_j$, where B is the bin in which t_i is located.

Drawbacks of scale based normalization

Scale based normalization relies on linearity between the measured peak intensity and the abundance of the molecule. In spectra with only one high abundant peak (which happens quite often), the rest of the peaks may be difficult to distinguish from background noise after normalization. In addition, when all peak intensities are low it may be difficult to distinguish the normalized values from each other.

7.2.2   Rank based normalization

Rank based normalization is used to overcome the drawbacks of scale based normalization. In this case

$$N(t_i) = \text{rank } t_i$$

with the highest intensity value getting rank one.

A possible variation is to divide the rank by the number of peaks in the spectrum:

$$N(t_i) = \frac{\text{rank } t_i}{n}. \tag{7.1}$$

For microarrays several statistics based on the differences in the ranks are defined for analyzing differential expression of genes across a set of chips, see for example Park et al. (2003a). If the value of the statistic is small, the gene is considered nondifferentially expressed.

7.2.3   Combining scale based and rank based normalization

Na and Paek (2006) proposed a method utilizing the best properties from relative and rank based normalization, called *cumulative intensity normalization*. This method normalizes a peak intensity value to the sum of all values with rank less or equal to the rank of the value under consideration, divided by the TIC:

$$N(t_i) = \frac{1}{\text{TIC}(T)} \sum_{\forall j: rank\ t_j \leq rank\ t_i} t_j.$$

Table 7.1   Example values for intensities and normalized intensities.

Normalizing type	Intensities									
Original	2.4	2.6	2.8	3.2	3.4	6.4	9.4	12.6	14.0	37.2
TIC	0.03	0.03	0.03	0.03	0.04	0.07	0.10	0.13	0.15	0.40
Cumulative	0.03	0.05	0.08	0.12	0.15	0.22	0.32	0.46	0.60	1.00

The result is a smoother curve of increasing normalized intensities, without the big leaps that are commonly seen when using scale based normalization methods.

Example Consider the intensities in Table 7.1, and the normalized values using TIC and cumulative intensity normalization respectively.
△

Figure 7.2 illustrates the normalization methods for a larger amount of data.

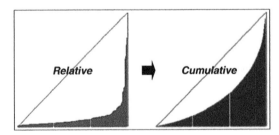

Figure 7.2   Illustrating scale based (in the figure called relative) and cumulative intensity normalization. Reprinted from Na and Paek (2006) by permission of American Chemical Society.

7.2.4   Reproducibility of the normalization methods

Degroeve et al. (2011) compared seven different normalization methods, where the strategy used for comparison was the argument that *a good normalization algorithm should increase the reproducibility of observed peak intensities for redundant MS/MS spectra*. Thus $N(t_i)$ should ideally be similar in all MS/MS spectra produced from the same peptide.

In order to assess reproducibility they used the *quartile coefficient of dispersion, qcod* statistic (see Chapter 20.7.2 for the definition of quartiles). Let Q_i be the quartiles for the distribution of the normalized intensities of a specific peak i across redundant spectra. Then the qcod is calculated as:

$$qcod(t_i) = \frac{Q_3 - Q_1}{Q_3 + Q_1}.$$ (7.2)

Thus for each normalization method $qcod(t_i)$ was calculated for each peak i. As a value for comparison the median across the peaks in a spectrum (least value best) was used. Based on an analysis of two unrelated datasets it was found that normalization by the vector norm was the best of the scale based methods. Also, the method in Equation 7.1 had the least median qcod of all the methods.

Example Assume that we have 12 spectra for a given peptide, and that there are $n = 10$ peaks in each the spectra. For a peak i the respective ranks in all the spectra are: 4, 5, 5, 3, 4, 5, 5, 7, 4, 3, 4, 4. Using Equation 7.1 we get for $N(t_i)$: 0.3, 0.3, 0.4, 0.4, 0.4, 0.4, 0.4, 0.5, 0.5, 0.5, 0.5, 0.7. Then $Q_1 = 0.4$, $Q_3 = 0.5$ and $qcod(t_i) = 0.11$. To find the median qcod we then calculate $qcod(t_i)$ for all ten peaks across the 12 spectra.

△

7.3   Normalization at the peptide and protein level

Depending on the quantification method used, methods for spectral normalization can in principle be reused for peptides and proteins. However, in most cases more advanced procedures are used for normalization at these levels. Note that log transformation of the abundances is typically carried out before normalization as this has been shown to give better results.

As an example consider two proteome profiles $P = \{P_1, P_2, \ldots, P_n\}$ and $Q = \{Q_1, Q_2, \ldots, Q_m\}$ (in many cases $n = m$). Note that, while our example concerns proteome profiles, these methods can also be used for normalizing peptide abundances.

When normalizing there are (at least) two questions that have to be answered:

1. Which proteins should be included in the normalization? Each protein with a measured abundance value in both samples (possibly limiting to those with a value above a specific threshold) could be included. Another alternative is to use specifically selected proteins. Three approaches for gene microarrays are mentioned in Park et al. (2003b): All genes on the array, constantly expressed genes, and control genes. Another option is to use the rank invariant genes.

2. Which normalization method should be used?

Note that the general task is to normalize over a set of samples. The most straightforward method is to select one of the samples as a reference sample, and then normalize all the others, one at a time, against this reference. This approach is used in the methods described in the following three sections. For the quantile normalization method however, normalization of several samples is performed as a single operation on all the samples together.

7.4   Normalizing using sum, mean, and median

When the total abundances are (approximately) equal in the two profiles, and the deviations (the differences between the real and measured values) for each protein depend linearly on the measured abundance, we can normalize using the sum. $K = \frac{\sum_{i=1}^{n} P_i}{\sum_{i=1}^{n} Q_i}$ is calculated, and each Q_i multiplied by K.

When the deviations for each protein are a constant, we can normalize using the mean. $K = \bar{P} - \bar{Q}$ is calculated, and K is added to each Q_i. Alternatively we can use the median instead of the mean, which is more robust against outliers.

7.5   MA-plot for normalization

In Chapter 6 we mentioned that log transformation is often used for statistical normalization, and this transformation can also be used in experimental normalization. Briefly the effect can be summarized as:

- the variation of log transformed abundances is less dependent on absolute magnitude;

- normalization is usually additive for log transformed intensities;

- taking logs evens out highly skewed distributions, that is, makes the distributions more normal.

For pairwise normalization one can plot the resulting values $(\log P_i, \log Q_i)$ in a scatter plot. However, it has been shown that another type of plot, called an *MA-plot*, is better for normalization purposes. The MA-plot is a plot with A (for average) and M (for minus) on the axes, where (using base 2):

$$A_i = \frac{\log_2 P_i + \log_2 Q_i}{2} = \frac{\log_2(P_i \cdot Q_i)}{2}; \quad M_i = \log_2 P_i - \log_2 Q_i$$
$$= \log_2 \frac{P_i}{Q_i}. \tag{7.3}$$

The reverse transformations to q_i, p_i are:

$$P_i = 2^{A_i + \frac{M_i}{2}}; \quad Q_i = 2^{A_i - \frac{M_i}{2}}. \tag{7.4}$$

The transformations involved in creating an MA-plot carry out a rotation and scaling of the $(\log_2 P, \log_2 Q)$ plot, as shown in Figure 7.3.

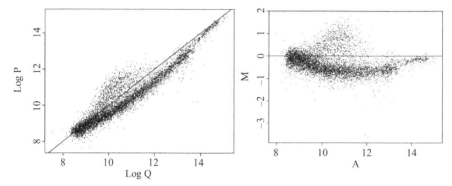

Figure 7.3   An illustration of a ($\log_2 P$, $\log_2 Q$) *plot (left), and the corresponding MA-plot (right). Reprinted from Dudoit et al. (2002) by permission of Statistica Sinica.*

As a result, trends (linear or nonlinear) due to shifts in the measured abundances will be more clearly visible in the MA-plot. Also note that the MA-plot is capable of showing shifts in both samples, not just in one.

From Figure 7.3 we can see that if the abundances are unchanged for the two samples, then under ideal conditions the points should be located on the $\log_2 P = \log_2 Q$ line for the left plot, and on the line $M = 0$ for the right plot.

Normalization is performed on the M-values, deriving normalized M'_i from the original M_i values. The normalized abundance values are calculated from the reverse transformation (replacing M by M' in Equation 7.4):

$$P'_i = 2^{A_i + \frac{M'_i}{2}}; \quad Q'_i = 2^{A_i - \frac{M'_i}{2}}. \tag{7.5}$$

The actual normalization can be performed in different ways, depending on our expectations of the systematic variations. In the following we describe two alternatives.

7.5.1   Global intensity normalization

Global intensity normalization is used when one suspects that there is a systematic fixed shift in one of the profiles compared to the other. The presence of a systematic shift can be investigated by considering the arithmetic mean, or the median, $\bar{M}$ from the M_is. A value of $\bar{M} = 0$ indicates that no shift is present, and normalization should therefore try to achieve an arithmetic mean (or median) of zero for the normalized values of M. This can be easily achieved by subtracting the mean (or median) $\bar{M}$ from each M_i: $M'_i = M_i - \bar{M}$. Note that if we now calculate the arithmetic mean of the normalized values, we always end up with a value of zero:

$$\frac{\sum M'_i}{n} = \frac{\sum (M_i - \bar{M})}{n} = \frac{\sum M_i - n\bar{M}}{n} = \bar{M} - \bar{M} = 0.$$

This means that we get a fixed shift in the M direction, effectively centering the abundance ratios for the normalization technique on the mean or the median. It is therefore also called *central tendency normalization*.

From Equation 7.5 we get:

$$P'_i = 2^{A_i + \frac{M'_i}{2}} = 2^{A_i + \frac{M_i - \bar{M}}{2}} = 2^{A_i + \frac{M_i}{2} - \frac{\bar{M}}{2}} = 2^{A_i + \frac{M_i}{2}} 2^{-\frac{\bar{M}}{2}} = P_i 2^{-\frac{\bar{M}}{2}}.$$

Using a similar derivation, we find:

$$Q'_i = Q_i 2^{\frac{\bar{M}}{2}}.$$

From these we calculate the ratio of P'_i over Q'_i as:

$$\frac{P'_i}{Q'_i} = \frac{P_i 2^{-\frac{\bar{M}}{2}}}{Q_i 2^{\frac{\bar{M}}{2}}} = \frac{P_i}{Q_i} 2^{-\bar{M}}.$$

The normalization value $2^{-\bar{M}}$ for the ratio is thus found from the MA-values.

Example Suppose we have $P_1 = 8$, $P_2 = 16$, $Q_1 = 4$, $Q_2 = 8$, then we find $A_1 = 2.5$, $A_2 = 3.5$, $M_1 = 1$, $M_2 = 1$. Furthermore, we find that $\bar{M} = 1$, and $M'_1 = M'_2 = 0$. From this we get $\frac{P'_1}{Q'_1} = 1$, and $\frac{P'_2}{Q'_2} = 1$.
△

7.5.2   Linear regression normalization

Linear regression (see Chapter 20.9) normalization can be used when one suspects that there is a systematic bias (shift) that depends linearly on the magnitude of the abundances. It can be performed by applying least squares regression to the MA-plot. The regression line is given by $M = k_1 A + k_2$.

Now consider a point (A_i, M_i) in the plot. Let $M^*_i = k_1 A_i + k_2$ be the predicted value calculated from the regression line. Then define the normalized value as $M'_i = M_i - M^*_i$. Note that a regression line with a positive slope indicates that the bias is increasing as the values of A_i increases.

7.6   Local regression normalization – LOWESS

Local regression normalization is frequently used for normalization when one suspects that there is a systematic bias that does not depend linearly on the magnitude of the abundances. *LOWESS – (Robust) LOcally WEighted regression and Smoothing Scatter plots* is the most common local regression analysis method.

Consider a $P - Q$ plot, where $P = P_1, P_2, \ldots, P_n$ and $Q = Q_1, Q_2, \ldots, Q_n$. New values $\hat{Q}_1, \ldots, \hat{Q}_n$ are then calculated, with the point $(P_i, \hat{Q}_i)$ called the smoothed point at P_i, and $\hat{Q}_i$ called the fitted value at P_i.

The procedure starts by first calculating initial values $\{\hat{Q}_i\}$, using (weighted) local regression of the nearest surrounding (on the P-axis) P, Q values, by a regression polynome. New values for $\{\hat{Q}_i\}$ are calculated iteratively, using newly calculated weights in each iteration. How many iterations to perform is specified at the start of the calculation. For further details see Cleveland (1979).

7.7 Quantile normalization

Quantile normalization is based on the assumption that the abundance distributions are similar for the different samples.

This method is used to normalize several samples in one operation, and has been shown to be useful for normalization of microarray data.

The assumption of similar distributions for the different samples essentially means that the abundances in all samples can be normalized to a common distribution $C_1, C_2, \ldots, C_n$. However, the ordering of the abundances in a sample should not change. This means that if the abundance of protein 1 in a sample is larger than the abundance of protein 2 in the same sample, the normalized abundance of protein 1 should remain larger than the normalized abundance of protein 2. A procedure for performing the normalization is described below.

Let there be n samples with m protein (or peptide) abundances in each sample. Further let

- $A_{i,j}$ be the abundance of protein i in sample j. Let A_j be all abundances from sample j, and A the table with $\{A_j\}$ as columns.

- $N_{i,j}$ be the rank of $A_{i,j}$ in sample j, and N the table of all $N_{i,j}$s.

The normalization is then described as:

1. Sort each column in A and form a table $\mathring{A}$ of all the sorted columns. Also determine $N_{i,j}$ during this operation.

2. For each row i calculate the mean $\mathring{A}_i$ in $\mathring{A}$, and form the column $\bar{A}$.

3. Achieve A normalized, A^*, by $A^*_{i,j} = \bar{A}_{N_{i,j}}$, we thus replace each value by a $\bar{A}_i$ value, where the rank i is determined based on the original rank $N_{i,j}$ of that value.

Example   Suppose that we have two samples $A_1 = \{2, 4, 7, 6, 3\}$ and $A_2 = \{1, 8, 5, 2, 6\}$. Even though we have only two samples, it should be clear that the same procedure can be used equally well for several samples. We can then form $A, N, \mathring{A}, \bar{A}, A^*$ as shown in the table below.

A		N		$\overset{\circ}{A}$		$\bar{A}$	A^*
2	1	1	1	2	1	1.5	1.5	1.5
4	8	3	5	3	2	2.5	4.5	7.5
7	5	5	3	4	5	4.5	7.5	4.5
6	2	4	2	6	6	6.0	6.0	2.5
3	6	2	4	7	8	7.5	2.5	6.0

Consider the original value $A_{3,1} = 7$. Its rank is $N_{3,1} = 5$. Based on this rank, we find $\bar{A}_5 = 7.5$ and we can thus normalize $A^*_{3,1} = \bar{A}_5 = 7.5$.

$\triangle$

To illustrate the effectiveness of this method we give another example.

Example Let the abundances in sample A_2 be double that of the corresponding abundances in sample A_1.

A		$\overset{\circ}{A}$		$\bar{A}$	A^*
2	4	2	4	3.0	3.0	3.0
4	8	3	6	4.5	6.0	6.0
7	14	4	8	6.0	10.5	10.5
6	12	6	12	9.0	9.0	9.0
3	6	7	14	10.5	4.5	4.5

After quantile normalization, the normalized values of the two samples become equal.

$\triangle$

7.8   Overfitting

Overfitting is an issue to consider for both statistical and experimental normalization, and means that the data is fitted too closely to the underlying assumption. If we consider relative abundances, the assumption is that most of the values should be one, that is, most proteins (or peptides) have unchanged abundances. The danger is then moving too much of the data towards one, making it practically impossible to discover differently abundant proteins. A concrete example involves the application of too many iterations in a LOWESS normalization.

Furthermore, overfitting can also have the opposite effect, leading to nondifferentially abundant proteins that are erroneously considered as differentially abundant.

7.9   Exercises

1. We have two MS-spectra with peak intensities (19, 29, 15, 16, 26, 14, 7) and (12, 19, 21, 8, 22, 5). The peaks with intensities 15 and 21 are found to correspond to the same peptide. We consider an intensity difference as significant if it has a fold change above 2. Check if the difference between the mentioned peaks is significant using:

   (a) maximum intensity;

   (b) TIC;

   (c) the square root of the squared intensities;

   (d) combined relative and rank.

2. Assume that we have two spectra containing three corresponding peaks of intensities $(T_1 : 4, 4, 4)$ $(T_2 : 2, 4, 6)$. Note that the TIC is the same for both. Show how the relative normalized intensities are when using TIC and using the vector norm. Comment on the result.

3. Explain that the formula in Equation 7.2 is reasonable for measuring the reproducibility of redundant MS/MS spectra.

4. We have the following abundances of two proteome profiles for six proteins: P: (10, 6, 8, 4, 7, 9) and Q: (5, 2, 8, 1, 3, 5). Perform normalization using TIC. Then, using MA-transformation find the differentially abundant proteins with a fold change greater than two.

5. We have three proteome profiles: (2, 6, 5, 10, 4), (5, 11, 4, 8, 9), and (3, 6, 7, 9,4). Perform quantile normalization.

7.10   Bibliographic notes

General normalization   Bolstad et al. (2003), Dudoit et al. (2002), Park et al. (2003b), Park et al. (2003a), Callister et al. (2006), Wang et al. (2008), Karpievitch et al. (2009b) Kultima et al. (2009).

Spectral normalization   Bern et al. (2004), Na and Paek (2006).

Comparing spectral normalizations   Norris et al. (2007), Degroeve et al. (2011), Deininger et al. (2011).

8

Statistical analysis

A sound statistical analysis of the acquired data is a critical step in obtaining reliable results from protein quantification experiments. In this chapter we describe statistics and statistical issues related to quantitative proteomics. If required, the necessary statistical background can be found in Chapter 20.

The basic question in quantitative proteomics can be formulated as: *Is protein P substantially differentially abundant in the two considered situations?* The answer is based on a statistical test saying 'yes' if the test is statistically significant, and 'no' if not. Which statistical test to use depends on the experimental setup and the data produced. We will consider two cases:

- replicate data for P are produced for the two situations;

- no (or a very limited amount of) replicate data for P is produced, but data for a large number of other proteins is also produced.

8.1 Use of replicates for statistical analysis

When replicates are available for the two situations and the data is normally distributed in each situation the T-statistic should be used. The T-statistic is however fairly robust against minor deviations from the normality assumption. For non-normal data the T-statistic will also be approximately correct if the number of replicates are large enough in both situations.

In the case where one has more than two situations where the data is normally distributed in each situation, analysis of variance (ANOVA) with an F-test should be used, see Chapter 20.5.2. The F-test is also fairly robust and gives approximately correct results from larger number of replicates in the non-normal case.

Computational and Statistical Methods for Protein Quantification by Mass Spectrometry,
First Edition. Ingvar Eidhammer, Harald Barsnes, Geir Egil Eide and Lennart Martens.
© 2013 John Wiley & Sons, Ltd. Published 2013 by John Wiley & Sons, Ltd.

For microarray data it has been observed that the T-test can put too much emphasis on the variance as compared to the difference between the means. Significance Analysis of Microarrays (*SAM*) is a variant of the T-test with an additional component in the denominator that includes the variance of the analyte over all replicates in the experiment, Tusher et al. (2001).

8.2   Using a set of proteins for statistical analysis

Often it is appropriate to perform experiments that do not produce replicate abundance data for P, but that do produce abundance data for a long list of other proteins. The assumption for the statistical analysis is, as explained in Chapter 5.11, that *only a limited number of proteins are differentially abundant in the two situations*. Several statistics/variables can be used for testing differential abundance given this assumption, of which we will describe three.

8.2.1 Z-variable

Let the calculated abundances of protein i in situation 1 and 2 be A_i, B_i respectively. For most of the proteins the ratios $\{R_i = \frac{A_i}{B_i}\}$ should then be one. Thus the ratios $\{R_i\}$ should form a distribution over calculated ratios from mainly not differentially abundant proteins. The ratio for our protein R_P, should then be tested to see where it is located within this distribution.

If we can assume an underlying normal distribution $N(\mu, \sigma)$ for the ratios R, we can use the Z-variable $Z = \frac{R-\mu}{\sigma}$. The Z-variable expresses the distance between a specific data point and the mean as the number of standard deviations. For the specific protein ratio R_P we can then calculate the probability that it is this many standard deviations removed from the mean (μ) under the assumption (or *null hypothesis*) that the measured protein is not differentially abundant between the situations.

From the table of the standard normal distributions we can find values for $z_{\alpha/2}$ such that $P(|Z| \geq z_{\alpha/2}) = \alpha$. Example values from such a table are:

α	0.1	0.05	0.02	0.01
$z_{\alpha/2}$	1.64	1.96	2.33	2.58

Values for the Z-variable can thus be used to directly estimate the likelihood of the observed abundances, a large value for protein P indicating a low likelihood of the protein P to not be differentially abundant. Since the actual population standard deviation is typically unknown, we can use the calculated sample standard deviation as an estimator, provided that the dataset contains a large enough number of data points to make the estimate sufficiently accurate.

Example Suppose that we have an abundance ratio R_P and we want to investigate if this ratio is unlikely large as determined for a normal distribution $N(\mu, \sigma)$. From the table above we have for a standard normal distributed variable Z that $P(|Z| \geq 1.96) = 0.05$. From this we can derive

$$P\left(\left|\frac{X - \mu}{\sigma}\right| \geq 1.96\right) = 0.05,$$

and further the expression

$$P(X \leq \mu - 1.96\sigma \text{ or } \mu + 1.96\sigma \leq X) = 0.05.$$

This means that if R_P is approximately two standard deviations or more away from the mean we can conclude that it is unlikely large and that P is differentially abundant.
△

Percentiles, see Chapter 20.7, can also be used as a robust way to determine the number of standard deviations a value is separated from the mean in a normal distribution. The connection between the number of standard deviations away from the mean and the percentiles is shown below for three numbers.

Standard deviations away from the mean	1	2	3
Percentiles	15.87	2.28	0.13
	84.13	97.72	99.87

This means, for example, that the 97.72th percentile is 2 standard deviations larger than the median. Note that the median is equal to the mean for normal distributions. Thus, if for a protein P we find an abundance ratio R_P we can calculate R_P's percentile and then estimate the value of the Z-variable, assuming the ratios are from a normal distribution.

Note that for using the procedure for the statistical analysis described in this subsection it is important that the protein P is determined before the ratios are calculated. It is thus, for example, incorrect to choose the protein with the lowest or highest ratios and analyze it in this way, see Section 8.5.

8.2.2   G-statistic

The G-statistic derives expected variable values from a large number of variable values (observations) based on a null hypothesis H_0. Let the observations be $\{o_1, o_2, \ldots, o_n\}$, and the expected variable values be $\{e_1, e_2, \ldots, e_n\}$ under the null hypothesis H_0 that we want to test. Then the G-statistic is calculated as

$$G = 2 \sum_{i=1}^{n} o_i \ln \frac{o_i}{e_i}. \tag{8.1}$$

Since the G-statistic is built around a logarithmic calculation, it can be approximated by the well-known chi-square statistic

$$Q = \sum_{i=1}^{n} \frac{(o_i - e_i)^2}{e_i}.$$

It is therefore possible to determine the significance of a value calculated for G through the chi-square distribution.

When considering the abundances of n proteins in two situations, and H_0 states that a specific protein is unchanged, we must:

- determine the variables;

- determine the (observed) values of the variables;

- calculate the expected variable values under H_0;

- calculate G;

- determine the degrees of freedom;

- use a chi-square distribution to obtain the significance.

Contingency table

When testing a protein P for being differentially abundant in situations S_1, S_2, under the assumption that only a few of the analyzed proteins are differentially abundant, it is appropriate to organize the abundance data in a contingency table. This is shown by Table 8.1, where the variables are:

- P_1, P_2 the abundances of protein P in situations S_1 and S_2, respectively;

- T_1, T_2 the total abundances of all the analyzed proteins (including protein P) in S_1 and S_2, respectively.

A contingency table is often used to record and analyze the relation between two or more categorical variables, here proteins and situations. The values in the last row and last column (except for the value in the lower right corner) are called *marginal totals*, and the value in the lower right corner is called the *grand total*.

Table 8.1   Contingency table.

Situations	P	All other proteins	All proteins
S_1	P_1	$T_1 - P_1$	T_1
S_2	P_2	$T_2 - P_2$	T_2
Total	$P_1 + P_2$	$(T_1 - P_1) + (T_2 - P_2)$	$T_1 + T_2$

Table 8.2   Spectral counts for 20 proteins in two situations.

Protein	1	2	3	4	5	6	7	8	9	10	
S_1	9	0	0	1	0	49	1	0	0	0	
S_2	5	1	3	1	2	13	0	9	4	8	
Protein	11	12	13	14	15	16	17	18	19	20	Sum
S_1	0	15	1	0	63	7	10	8	0	4	168
S_2	1	8	0	1	57	14	12	17	4	16	176

Table 8.3   Contingency table for testing protein 1 using the G-statistic.

Situations	Protein 1	All other proteins	All proteins
S_1	9	159	168
S_2	5	171	176
Total	14	330	344

Example   Assume that we have measured the abundance by spectral counts (see Chapter 14) and obtained the values for 20 proteins in two situations S_1, S_2, as shown in Table 8.2.[1]

Note that in this example we consider 0 as a real value, and not as a missing value.

We want to test whether protein 1 is differentially abundant at a significance level of 0.05. Given that we assume that most of the proteins are not differentially abundant, we must investigate whether protein 1 differs significantly from the rest. We therefore present the data in a contingency table as shown in Table 8.3.

From this table we calculate the expected number of spectral counts for protein 1 in the two situations under the null hypothesis of no change.

There are two variables, representing the measured abundances of protein 1 in the two situations. Thus $o_1 = 9$, $o_2 = 5$. The analysis is then performed as follows:

- The fraction of the total number of spectral counts in situation S_1 is $168/344 = 0.4884$, and for situation S_2 the fraction is 0.5116.

- Under the null hypothesis of no change, we can use these fractions to calculate the expected number of the 14 observed spectral counts for protein 1 to occur in situation S_1 and S_2, respectively: $e_1 = 14 \cdot 0.4884 = 6.8376$, $e_2 = 7.1624$.

- From Equation 8.1 we find $G = 1.3521$.

---

[1] The values are from an example of Bingwen Lu: http://www.lubw.com.

- There is one degree of freedom here (when the marginal values are fixed and we change one of the inner values, the other three are determined).

- From a chi-square distribution we find that for rejecting H_0 on the 0.05 confidence level, G should be at least 3.84, so protein 1 is not significantly differentially abundant between situations S_1 and S_2 according to this test.

Note that we do not need an explicit normalization for performing this test, since a global normalization is included implicitly in the test.

△

8.2.3   Fisher–Irwin exact test

When the number of data points is low, the G-test is not accurate enough. An alternative in such situations is the Fisher–Irwin exact test, based on calculating exact probabilities.

We again organize the data in a contingency table, and the assumption is (as before) that most of the proteins are not differentially abundant in the two situations. We will test if protein P is differentially abundant between situations (S_1, S_2). For simplicity we assume that all values in the table are integers.

The principle behind the test is to fix the values in the last row and last column (the marginal numbers). Our mathematical null hypothesis is then related to the probability of achieving the other four numbers, given the fixed marginal numbers. Note that as soon as one of the inner numbers is determined, all others are also determined.

The problem can be formulated as follows: When we know that $P_1 + P_2$ of the $T_1 + T_2$ are from protein P, and that T_1 of these $T_1 + T_2$ are from situation S_1, what is the probability that these $P_1 + P_2$ values would be distributed over S_1 and S_2 as shown (P_1, P_2)? Fisher showed that this probability is:

$$p = \frac{\binom{P_1 + P_2}{P_1}\binom{(T_1 - P_1) + (T_2 - P_2)}{T_1 - P_1}}{\binom{T_1 + T_2}{T_1}}.$$

From this we get:

$$p = \frac{T_1!T_2!(P_1 + P_2)!(T_1 - P_1) + (T_2 - P_2)!}{(T_1 + T_2)!P_1!P_2!(T_1 - P_1)!(T_2 - P_2)!}.$$

In hypothesis testing we calculate the probability of achieving the observed results *or more extreme results*. If the abundance of P is higher in S_1 than in S_2 (relative to the totals), we therefore use the alternative hypothesis that P is more abundant in S_1 relative to S_2. This means that we have a one-sided alternative, hence the more extreme result is that the value observed for P in S_1 is $P_1 + 1, P_1 + 2, \ldots$ We therefore also have to calculate the probabilities for these results, and use the sum of all the probabilities to test for significance. *Fisher has shown that it is enough to consider only the more extreme cases where the totals at the marginals remain fixed.* Note however that the number of values that have to be calculated can often be very large. Fortunately, computers make it easy to perform such calculations.

Example Consider the example in Section 8.2.2, for which the data is shown in Table 8.3, where P is protein 1. The probability of randomly observing the given distribution $(9, 5)$ is:

$$\frac{\binom{14}{9}\binom{330}{159}}{\binom{344}{168}}.$$

The more extreme values to be tested with the one-sided test are the values where P in S_1 is 10, 11, 12, 13, or 14. For P in $S_1 = 10$, the probability would for instance be:

$$\frac{\binom{14}{10}\binom{330}{158}}{\binom{344}{168}}.$$

△

8.3 Missing values

Missing values can occur at both the peptide and the protein level in protein quantification. The model we use for discussing missing values consists of two situations A and B and a number of replicates for each of these, $A_1, A_2, \ldots, A_n$ and $B_1, B_2, \ldots, B_m$.

Missing values occur when an abundance for a given protein is determined for some of the replicates but not for all. Given that such missing values can interfere with the statistical analysis, it is important to handle these correctly. Various factors can contribute to the absence or presence of a measurement in a replicate, of which the most important are discussed below.

8.3.1 Reasons for missing values

The reasons for missing values can be divided into *biological reasons* and *experimental reasons*, and this division is of crucial importance, as will be explained below.

Biological reasons: There are two common biological reasons for missing values:

- The protein is effectively not present in the sample.

- The protein abundance is too low to be detected, that is, the protein exists below the *detection limit*.

Experimental reasons: The protein exists in the sample in a measurable quantity, but is not detected. While accurate and careful execution of the experiment will reduce the number of such experimentally missing values, quite a few proteins will still be missed in a typical high-throughput experiment. There are several experimental reasons for missing values:

- The protein 'disappears' from the sample before the LC-MS/MS process, for example, due to precipitation during sample extraction or preparation, adsorption losses to the walls of the reaction vessels or instrument tubing used, or degradation due to remaining activity of proteases in the lysate. In general, the chance of losing a protein during sample handling goes up with each manipulation, and the number of sample processing steps is therefore often minimized.

- Peptides derived from the protein are not selected for MS/MS, or too few unique peptides are selected for the protein. This is a consequence of *under-sampling*, in which the mass spectrometer is flooded with more ions than it can analyze in a given time frame, and therefore resorts to a (stochastic) sampling of the presented analytes.

- The MS/MS spectra produced for the (unique) peptides derived from the protein are not identified with high enough confidence.

- In relative quantification, if one of the variant signals is close to the signal-to-noise threshold in that particular spectrum, it may be considered as noise.

It is important to note that while experimentally missing values are undesirable, biologically missing values are essential for the statistical analysis. It is therefore crucial to try and identify the reasons for the missing values.

- If a protein is missing in all replicates of a situation, it is likely due to a biological reason.

- If a protein is identified in only a few of the replicates, the situation becomes more complex, as it could be derived from biological reasons, experimental reasons, or both biological and experimental reasons.

If a complex situation of missing values is encountered, it should be investigated whether there is a pattern for where the missing values occur. This may give an indication of the underlying reasons for the missing values, and might also unveil experimental weaknesses. Examples include consistently low signals that can be observed upon manual inspection of the corresponding spectra, or the predominant absence of signal for many proteins in a given replicate, indicating problems with the replicate. Effects can be more subtle as well, including for instance the absence of hydrophobic peptides in general, leading to missing values for predominantly hydrophobic proteins.

Additionally, if only a few values are found for a given variant across the replicates, it is worth investigating whether these measurements are correct; indeed, a few incorrect measurements may resemble measurements with many missing values.

8.3.2   Handling missing values

When we do not know the reason for the missing values they still have to be treated in a reasonable way. Since the number of missing values in mass spectrometry based proteomics experiments is substantial, it is important to handle them correctly. There are at least three different ways to handle missing values:

- Ignore proteins with missing values. This approach is only useful if the number of proteins with missing values is (very) low.

- Perform the analysis based on the known (nonmissing) values, ignoring the missing values. This can be done if the number of known values is large enough for the analysis method used, for example, for a T-test at least two members in each group is required.

- Replace the missing values by an *imputed* value. However, note that determining reasonable imputed values is more difficult in mass spectrometry proteomics than in microarray experiments and in proteomics using 2D gels. The reason for this is that the analysis process performed to arrive at the original values is more complicated.

  Imputed values can be determined in several ways, where the most common are:

  – Choose a minimum value to replace the missing value. There are a number of different ways to choose the new value. Examples: Set it to zero, set it to the minimum nonzero value in the replicates for the situation, set it to a value randomly chosen among the lowest values. Note that choosing a minimum value assumes that the reason behind the missing values is the low abundance or absence of the corresponding proteins.

  – Choose an average or median value for the protein. The most reasonable calculation relies on the known values for the same protein in the same situation. Note that this imputation strategy assumes that the missing value is due to experimental reasons. Note also that this can effect the statistical test. However, it can be used almost harmlessly in tasks such as clustering and discriminant analysis.

- Missing values can also simply be maintained as missing values, especially if there is a biological reason for their absence, for example, a given protein is consistently absent from a particular patient.

8.4   Prediction and hypothesis testing

In large scale proteomics we simultaneously consider the abundances of a large number of proteins, and the tests for differentially abundant proteins have to be performed with this parallel testing in mind, as is described in Section 8.5. The tests are described in the context of predictions and hypothesis testing.

Table 8.4   Example data illustrating errors in prediction and hypothesis testing. DA means differentially abundant proteins.

		Predicted		
		DA	non-DA	Total
	DA	TP = 40	FN = 10	50
True	non-DA	FP = 30	TN = 920	950
	Total	70	930	1000

The type of prediction we consider can be described as a variant of the classification problem: The predicted results belonging to one class are called *positives*, while those predicted to belong to the other class are called *negatives*. In the specific case of differentially abundant proteins, the class of positive results comprises those proteins that are differentially abundant, while the class of negative results contains the proteins that are not differentially abundant.

8.4.1   Prediction errors

In prediction one can make errors. Four important terms are defined in this context:

TP – True Positives: The number of results predicted to be positives that are indeed positives. In our context this is the number of proteins that are predicted to be differentially abundant and that really show a difference in abundance between the situations.

FP – False Positives: The number of results predicted to be positives that are in fact negatives.

TN – True Negatives: The number of results predicted to be negatives that really are negatives.

FN - False Negatives: The number of results predicted to be negatives that are in fact positives.

We use the above terms to define some additional concepts that are illustrated using the data in Table 8.4, where we have both predicted and correct results. For example, from the table we see that there are 50 proteins that are really differentially abundant. At the same time, of the 70 that are predicted to be differentially abundant, only 40 are actually differentially abundant.

FPR – False Positive Rate: The proportion of actual negatives that are classified as positives. In our context, the FPR is the ratio of the number of wrongly predicted differentially abundant proteins to the number of not differentially abundant proteins. In Table 8.4 we find:

$$FPR = FP/(FP + TN) = 30/950 \approx 3\%.$$

FDR – False Discovery Rate: The proportion of predicted positives that are falsely predicted. In our context it is the ratio of the number of wrongly predicted differentially abundant proteins to the total number of predicted differentially abundant proteins. In Table 8.4 we find:

$$FDR = FP/(FP + TP) = 30/70 \approx 43\%.$$

FNR – False Negative Rate: the proportion of actual positives that are classified as negatives. In our context it is the ratio of the number of wrongly predicted nondifferentially abundant proteins to the number of differentially abundant proteins. In Table 8.4 we find:

$$FNR = FN/(FN + TP) = 10/50 = 20\%.$$

8.4.2 Hypothesis testing

Prediction is often performed by hypothesis testing, as detailed in Chapter 20.3.

In our quantitative proteomics context we first consider only one specific protein P. The procedure for this is briefly:

1. Formulate the null hypothesis H_0: The protein P is not differentially abundant in the two situations, meaning that any difference in abundance is due to chance, to noise, or to other experimental irregularities. The alternative hypothesis is that P is differentially abundant.

2. Define the desired significance level for H_0, for example 0.05.

3. Calculate some value V for the difference in abundance of P between the two situations.

4. Determine the probability for achieving the result V or higher, given H_0, and compare this probability to the significance level. If the probability is less than or equal to the significance level, conclude that P is differentially abundant in the two situations. If it is greater than the significance level, conclude that P is not differentially abundant.

When testing a long list of proteins, we can now quantify the overall frequency of the two types of error that can occur as explained below.

Type I error: Rejecting H_0 when H_0 is true. The probability of a Type I error is the significance level α, and the FPR is a quantification of it. (Note however that the correct results are typically not known, so the FPR can only be estimated rather than calculated directly.)

The *specificity* of a test is defined as $1 - \alpha$. For Table 8.4 this yields

$$\text{specificity} = 1 - FPR = TN/(FP + TN) = 920/950 = 96.8\,\%.$$

Specificity thus provides the proportion of correctly predicted nondifferentially abundant proteins among the total number of actual nondifferentially abundant proteins. Note again that this value has to be estimated, since the actual numbers are typically unknown.

Type II error: Accepting H_0 when H_0 is false. The probability of a Type II error is often denoted by β, and the FNR is a quantification of it.

The *power* of a statistical test is defined as the probability that it will reject H_0 when H_0 is false, or, in other words, the probability that it will not make a Type II error. Note that this means that the power is $1 - \beta$. For Table 8.4 this yields

$$\text{power} = 1 - FNR = TP/(TP + FN) = 40/50 = 80\,\%.$$

Power is therefore the proportion of correctly predicted differentially abundant proteins to the actual number of differentially abundant proteins. Because of this, the power is also sometimes denoted as the *sensitivity* of the test.

8.5   Statistical significance for multiple testing

In large scale proteomics we simultaneously consider the abundances of a large number of proteins, and often perform some sort of statistical test for each of these proteins, for instance a T-test. However, many of these tests were not created with this sort of parallelism in mind, and their statistical significance is therefore based on the assumption that the test is only performed once.

For example, for a T-test with $\alpha = 0.01$, there is a 1 % chance of getting a Type I error. This may be tolerable when performing a single test, but when employing this test hundreds or thousands of times (simultaneously), the probability of getting a false positive result in any one of these tests will increase dramatically, and we can therefore expect that several of the proteins will test below the significance level just by chance.

Example Suppose that we have two situations for which we want to find differentially abundant proteins. $\alpha = 0.01$, and we test 200 proteins simultaneously. For each test there is a 1 % chance of making a Type I error and thus obtaining a significant result even though the protein is in fact not differentially abundant between the two situations. However, because we run 200 tests (one for each protein), each of which has a 1 % chance of providing a false positive, we can expect $200 \cdot 0.01 = 2$ false positive results among the 200 proteins.

$\triangle$

Table 8.5   Table illustrating FWER. DA means differentially abundant. See text for explanation.

		\multicolumn{2}{c}{Test results}		
		DA	non-DA	Total
	DA	TP	FN	$n - n_0$
True	non-DA	FP	TN	n_0
	Total	D	n-D	n

The above example illustrates that our methods for comparing means cannot be directly used for multiple testing, and various solutions have been proposed for dealing with this so-called multiple testing problem.

8.5.1   False positive rate control

False positive rate control means that the threshold for significance for the p-values is adjusted. The name of this method is based on the fact that the FPR is a quantification of the p-value. Adjustment is performed by considering all p-values together, hence the alternative name *family-wise error-rate control (FWER)*.

This method calculates the probability of making at least one false positive call (one Type I error, hence FPR control) among all hypothesis tests when performing multiple testing. Consider contingency Table 8.5 for the explanation.

We have n proteins, which means n hypothesis tests. DA means differentially abundant, n_0 is the actual number of nondifferentially abundant proteins, and D is the number of predicted differentially abundant proteins. FP is the number of Type I errors, and FN the number of Type II errors. FWER is defined as

$$FWER = Pr(FP \geq 1),$$

which is the probability that there is at least one false positive test result.

We first assume that n is known, but n_0 is unknown. In statistical terms we consider D as an observable random variable, and TP, FN, FP, TN as unobservable random variables.

The significance level for each individual hypothesis test is called the *nominal significance level*. We can calculate the significance level associated with the FWER by increasing the requirement we set for rejecting H_0 for each individual test.

For the following we assume:

- we have n tests (proteins);
- the tests are sorted in increasing order of the p-values, p_1 being the smallest and p_n the largest;
- the nominal significance threshold is α.

Table 8.6   A simple example of values for Bonferroni and Holm–Bonferroni corrections. $n = 100$, and the four most significant p-values are shown.

	p_i	Bonferroni $p_i n$	Holm–Bonferroni $p_i(n - i + 1)$
1	0.00002	0.002	0.002
2	0.00039	0.039	0.039
3	0.00051	0.051	0.050
4	0.00106	0.106	0.103

Bonferroni correction

The simplest technique is called Bonferroni correction. The threshold for Bonferroni correction is achieved by dividing the nominal significance threshold by the number of tests. This means that H_0 is rejected for test i if

$$p_i \leq \frac{\alpha}{n} \quad (\text{or } p_i \cdot n \leq \alpha).$$

Holm–Bonferroni correction

Bonferroni correction is rather conservative, erring on the side of caution, and thus yielding many false negatives. The *Holm–Bonferroni correction* is a step-wise variation that makes the method less conservative (hence giving it more power). H_0 is now rejected for a test i if

$$\forall j \leq i : p_j \leq \frac{\alpha}{n - j + 1} \quad (\text{or } p_j \cdot (n - j + 1) \leq \alpha).$$

Note that this is a step-wise procedure, starting from $i = 1$ and progressing until the first test that no longer rejects H_0 is reached. Also note that this means that there might be an i for which $p_i \leq \frac{\alpha}{n-i+1}$ that will not reject the null hypothesis, because there has already been a $k < i$ for which $p_k > \frac{\alpha}{n-k+1}$.

Example Based on the sorted p-values from Table 8.6, and $\alpha = 0.05$ and $n = 100$, Bonferroni correction retains only two of the test results as significant ($p_i n \leq 0.05$), despite the fact that the original p-values of the individual tests were all below the nominal significance threshold. Using Holm–Bonferroni correction yields three significant p-values.
△

8.5.2   False discovery rate control

For microarray experiments it has been shown that error rate control based on false discovery rate yields greater power than FWER, but also that this comes with an increasing likelihood of producing Type I errors.

A list of scores for the proteins is presented, sorted in an order such that the first protein in the list has the highest probability for being differentially abundant. The scores could be p-values, but need not be. Remember that FDR is estimated as:

$$FDR = FP/(FP + TP).$$

If we consider as positives (differentially abundant) the i first tests in our sorted score list, then $FP + TP = i$, and

$$FDR_i = \frac{FP_i}{i},$$

where the index specifies that the first i proteins in the list are considered positives.

We now have to estimate FP_i, the number of predicted positives that are actually false. This estimation is not straightforward however, and can be performed in various ways, of which we here describe two.

Benjamini and Hochberg correction

We here assume that the scores are p-values. Benjamini and Hochberg correction estimates FP by considering that

$$FPR = FP/(FP + TN)$$

is an approximation of the p-value. We can thus write

$$p_i \approx FP_i/(FP_i + TN_i) \implies FP_i \approx p_i \cdot (FP_i + TN_i).$$

$FP_i + TN_i$ is unknown, but for large n with most of the tests actually negative, that is, most proteins are not differentially abundant, it can be approximated by n. For the list of the first i we therefore simply form

$$FDR_i = \frac{FP_i}{i} \approx \frac{p_i \cdot n}{i}.$$

For each list of type $\{p_1, \ldots, p_i\}$ we could compare FDR_i to a threshold, and return the lists as significant when FDR_i is less than or equal to the threshold. However, FDR_i is not an increasing function of i, which is unwieldy in this context. In order to obtain an increasing function, q-values are defined as:

$$q_i = min_{j:i \leq j \leq n} FDR_j. \tag{8.2}$$

These q_is increase with i, and they can therefore be unambiguously compared to a defined q-threshold. The procedure is to first calculate q_n, and then in an iteration calculate $q_i = \min(FDR_i, q_{i+1})$.

Table 8.7 A simple set of example values for Benjamini and Hochberg correction. The six most significant p-values are shown, and $n = 600$ is used.

| | p_i | Benjamini and Hochberg | |
		$\frac{p_i n}{i}$	q_i
1	0.00002	0.012	0.012
2	0.00007	0.021	0.018
3	0.00009	0.018	0.018
4	0.00020	0.030	0.030
5	0.00028	0.034	0.034
6	0.00036	0.036	0.036

Example Table 8.7 shows Benjamini and Hochberg corrections along with the corresponding q-values for some example p-values. Note that q_2 is different from the corresponding Benjamini and Hochberg values to ensure an increasing function. $\triangle$

Storey and Tibshirani

Storey and Tibshirani (2003) has performed a formal analysis of the q-values. The details are outside the scope of this book, but the ideas are used to show how one can estimate the FDR for proteomics experiments.

We consider

$$FDR_i = \frac{FP_i}{i},$$

and assume the list of scores $\{s_j\}$ are in increasing order. Also n_0 (unknown) of the n tested proteins are not differentially abundant.

Estimating FP_i FP_i is the number of the n_0 nondifferentially abundant proteins with scores $\leq s_i$, and are thus incorrectly considered positives. We then make some additional assumptions:

- we have a score distribution of r (large) random nondifferentially abundant proteins;

- f_i of them have score $\leq s_i$;

- if n_0 is large we can assume that the n_0 nondifferentially abundant proteins have a similar distribution to the distribution of the r proteins above.

Thus we derive

$$\frac{FP_i}{n_0} = \frac{f_i}{r} \Rightarrow FP_i = \frac{f_i n_0}{r},$$

and from this

$$FDR_i = \frac{FP_i}{i} = \frac{f_i n_0}{ri} \approx \frac{f_i n}{ri}, \tag{8.3}$$

where we have approximated n_0 by n, which should be reasonable when n (and hence n_0) is large. The q-values are calculated as in Equation 8.2.

One assumption mentioned above is that we have a score distribution of nondifferentially abundant proteins. Next we will therefore discuss how to obtain such a score distribution.

Constructing a random distribution   To construct a score distribution of non-differentially abundant proteins, we can use a general bioinformatics technique that generates such a distribution from real data, most often by the use of permutations.

Consider a quantification experiment of n proteins for two situations, each having m replicates that are used in the statistical analysis. The resulting data thus consists of n profiles (for the n proteins), each consisting of $2m$ abundance values (one set of replicate values for each situation). We want to generate a set of random profiles corresponding to nondifferentially abundant proteins, and calculate how many of them that score less than or equal to a threshold t. These random profiles should ideally be typical profiles that can be produced by the experiment for nondifferentially abundant proteins. An assumption is that such random profiles can be constructed by permutation of the real profiles obtained in the experiment.

We first consider one profile. Note that permuting values inside only one situation will not change the score. In order to calculate the number of permutations we can achieve for each profile, we need to consider the number of ways one can draw m values from a set of $2m$ values when the order of the numbers does not matter and the numbers are not replaced. This is given by:

$$\binom{2m}{m} = \frac{(2m)!}{m!m!}.$$

Also, since completely swapping the profiles between the two situations also results in the same score (for a two-sided test), the number of permutations for one profile (including the original pattern) is further reduced to:

$$\frac{(2m)!}{2m!m!}.$$

It is reasonable to believe that permuting profiles from nondifferentially abundant proteins will result in profiles that can be achieved from nondifferentially abundant proteins. For profiles from differentially abundant proteins this assumption may be considered more doubtful. However, assuming that the number of differentially abundant proteins is low they should not disturb the main distribution very much.

One can also reasonably assume that the differentially abundant proteins will occur in the first h proteins in the sorted list, and we can simply avoid these in the permutation.

The problem of unknown n   In microarray experiments a fixed number of genes are investigated (defined by the number of probes on the chip), thus n is known. For a large scale quantitative proteomics experiment however, the number of investigated proteins is essentially unknown. Thus n in Equation 8.3 is unknown.

In protein quantification experiments only a number n_p, of the proteins in the samples are identified and quantified. However, if n_p is large enough, and if it is reasonable to believe that these n_p proteins are a representative selection of all the proteins in the samples, n_p can be used as a proxy for n. Representative selection here means that the identification and quantification of proteins does not depend on whether they are differentially abundant or not.

Whether the criterion of representative selection is valid or not has to be examined for each quantitative method used.

8.6   Exercises

1. We have calculated the logarithmic abundance ratios for a set of proteins, and the ratio for a specific protein is $x_P = 0.62$. The mean and standard deviation for the distribution of all abundances are $\mu = 0.02$ and $\sigma = 0.27$. We assume that data are from a normally distributed population. Use the Z-variable to find out if the considered protein is differentially abundant when using a confidence level of 98 %.

2. Use the G-statistic to test if protein 6 in Table 8.2 is differentially expressed at the 0.05 confidence level.

3. We have an experiment using replicates. Explain what effect using an imputed value for a missing protein value has on a T-test if:
   (a) the imputed value is calculated as the average of the protein values in the other replicates;

   (b) the imputed value is determined by another independent method (for example using mRNA abundances).

4. We have performed an experiment to test a procedure for determining differentially abundant proteins. We tested 400 proteins for which we know the (correct) values, and got the following results: $TP = 20, FN = 8, FP = 5, TN = 367$. Calculate FPR, FDR, FNR, specificity and power.

5. Assume that we have tested 8 proteins, and got the following p-values:

$$0.001, \ 0.004, \ 0.008, \ 0.010, \ 0.017, \ 0.080, \ 0.120, \ 0.340.$$

Let $\alpha = 0.05$.

(a) Find the proteins that are assumed to be differentially expressed. First by using Bonferroni correction, and then by using Holm–Bonferroni corrections.

(b) Calculate q_i-values for Benjamini and Hochberg correction, and find the longest significant list of proteins when the threshold for the q-value is 0.02. (Note that this is not a realistic example since the number of proteins tested is far too small.)

8.7   Bibliographic notes

Missing values   Pedreschi et al. (2008), Albrecht et al. (2010), Karpievitch et al. (2009a), Karpievitch et al. (2009b), Li et al. (2011).

Multiple testing   Benjamini and Hochberg (1995), Benjamini and Yekutieli (2001), Storey and Tibshirani (2003), Käll et al. (2009), van Iterson et al. (2010).

9

Label based quantification

One of the main distinguishing features between quantification approaches is the use of labeling. In label free approaches, separate quantification runs are performed for each sample, whereas the label based approaches can analyze two or more samples simultaneously in a single quantification run. Because of the costs involved in labeling, label free approaches provide the cheapest option, but they also incur the most complex downstream analysis. We will therefore first cover the slightly simpler data analysis of the label based approaches.

9.1 Labeling techniques for label based quantification

In label based quantification one typically examines two or more protein samples simultaneously in one quantification run. Most often the samples are from more than one situation.

To distinguish the variants of the peptides, the peptides are tagged with different labels, where each sample gets a unique label. After labeling the samples are combined, providing a single mixture for analysis. In some labeling methods one of the samples can simply remain unlabeled, allowing $n + 1$ samples to be combined, with n being the number of labels.

Note that this procedure does not cover the use of labeling in targeted or absolute quantification, where synthetically produced, labeled standards are used, and separate runs are performed to compare each sample to the synthetic standard. The following therefore concerns discovery oriented quantification experiments. Targeted quantification is discussed in Chapter 15, and absolute quantification in Chapter 16.

Computational and Statistical Methods for Protein Quantification by Mass Spectrometry,
First Edition. Ingvar Eidhammer, Harald Barsnes, Geir Egil Eide and Lennart Martens.
© 2013 John Wiley & Sons, Ltd. Published 2013 by John Wiley & Sons, Ltd.

9.2   Label requirements

The labels ought to satisfy the following criteria:

- The different variants of a peptide should have the same ionization characteristics, that is, the relative peptide variant ion currents should mirror the relative peptide variant abundances in the combined sample.

- The different variants of a peptide should have the same chromatographic characteristics, that is, the peptide variants should elute from the chromatographic column during the same time interval (equal retention time). This assures that the variants occur in the same spectra.

- The labeling should not complicate the MS/MS spectra too much, that is, they should not interfere with peptide identification.

- The treatment of the spectra is easier if there is one, and only one, label for each peptide variant. This makes the task of discovering pairs of peptide variants easier. However, this last condition is only truly satisfied for some of the labeling techniques.

The requirements above are best satisfied by using stable isotopes. A stable isotope is an isotope that is not radioactive, meaning that the nucleus is stable and does not decay. This ensures a constant quantity of the label (as it does not decay away over time), and makes the label safer to handle, as it emits no radiation. Because the different isotopes of an element all have equal numbers of protons and electrons, the different variants of a peptide have the same chemical properties. Due to different numbers of neutrons, however, they do have different masses, thus occurring at different positions in the MS or MS/MS spectra. It is this latter difference between isotopes in physical properties that allows the samples to be separated by a mass spectrometer.

It should be noted, however, that the difference in physical properties extends to the behavior of a peptide on an LC system, causing a potential shift in retention time between different variants. This effect is particularly prominent for deuterium, the heavy stable isotope of hydrogen. Larger elements (such as nitrogen or carbon) do not display noticeable shifts between their different stable isotopes, and are therefore generally preferred in labels over deuterium.

9.3   Labels and labeling properties

There are several ways to classify the various methods for labeled protein quantification.

9.3.1   Quantification level

Existing labeling techniques can be divided into two groups by considering the MS level that is used for determining protein abundance, either at the MS level, or at the MS/MS level. The two groups use different methods for determining relative peptide abundances, and this will be described in detail in the next chapters. Note that in both cases, MS/MS spectra are used for peptide identification.

9.3.2 Label incorporation

There are essentially three techniques available for incorporating labels into peptides or proteins: Metabolically, chemically, or enzymatically.

Metabolic labeling The labels are incorporated directly into living cells by growing the cells in media that contain stable isotopes. Stable isotopes are usually introduced in the medium in the form of molecules that the cells require to grow; in the case of proteomics these tend to be essential amino acids or amino acid precursors. The labeled molecules are then taken up by the cell, and incorporated into newly produced macromolecules through growth and turnover. In proteomics, these macromolecules are of course proteins.

For two-way comparisons two cell cultures are grown, one in a light medium (incorporating the light isotope) and one in a heavy medium (incorporating the heavy isotope). As a result, proteins in one cell culture will carry the light label, and proteins in the other cell culture will incorporate the heavy label. Note that the light label in these cases tends to be the most abundant natural isotope, and thus is not really a label at all.

One advantage of incorporating the labels directly into living cells is that the accuracy of the quantification will not be affected by any differences in the subsequent digestion and purification steps, since the samples can be mixed directly after the harvesting of the cells, and they will thus be subjected to all subsequent *in vitro* steps as a single mixture.

Chemical labeling A labeled chemical is covalently bound to the molecules to be quantified. The chemical is then incorporated into the complete proteins prior to digestion, or into the peptides after digestion.

Enzymatic labeling The labels are incorporated by the protease during digestion, obviating the need for additional protocol steps in the quantification procedure.

Figure 9.1 illustrates the different ways in which labels can be incorporated.

Synthetic and chemically labeled peptides can also be used as *internal standards* added to the sample during the *in vitro* processing. By spiking known quantities of such peptides into the unlabeled samples, absolute quantification of peptides and proteins can be achieved in small samples. The use of synthetic peptides is further described in Chapters 15 and 16.

9.3.3 Incorporation level

The labels are incorporated either at the protein or at the peptide level. Labels incorporated at the protein level should of course remain attached to the peptides after digestion.

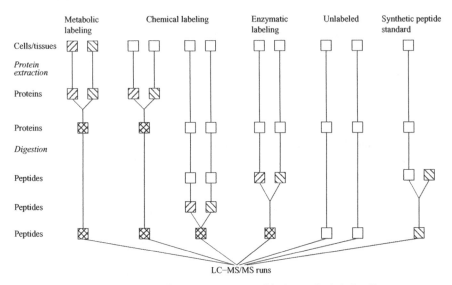

Figure 9.1 Different ways of incorporating stable isotopic labels. Empty squares represent unlabeled samples, while hatched squares represent labeled samples. For comparison the label free approach and the use of a synthetic, labeled standard are also illustrated.

9.3.4   Number of compared samples

Some of the techniques only allow the comparison of two samples in an LC-MS/MS run, while others allow a broader multiplexing.

9.3.5   Common labels

Table 9.1 shows the most commonly used labels. In the following chapters these approaches will be explained in more detail.

9.4   Experimental requirements

In addition to the assumptions listed in Chapter 5.11 some additional assumptions are employed for label based methods:

- The labeling process is such that for a specific peptide or protein an equal percentage of the molecules in the sample is labeled for each variant. This assumption can be compromised if the light and heavy label have different reactivities. In most cases however, this issue does not arise due to the identical chemical behavior of the light and heavy isotopes, and by extending the time spent on the labeling process, complete labeling can (in most cases) be ensured.

Table 9.1   The most commonly used labels for protein quantification.

	Quantification level	Incorporation	Incorporation level	Number of samples
SILAC	MS	Metabolically	Protein	2–3
ICAT	MS	Chemically	Protein	2
ICPL	MS	Chemically	Protein	2–4
Dimethyl labeling	MS	Chemically	Peptide	2–4
^{18}O	MS	Enzymatically	Peptide	2
iTRAQ	MS/MS-Reporter	Chemically	Peptide	4, 8
TMT	MS/MS-Reporter	Chemically	Peptide	2, 6
IPTL	MS/MS-Fragment	Chemically	Peptide	2

- The final mixture is composed of an equal amount of protein or peptide extracts from each of the samples to be combined. Note that including too much (or too little) of one particular extract will cause a systematic shift that it is possible to compensate for in the data analysis.

In order for these assumptions to be fulfilled, detailed protocols must be accurately followed and care must be taken when combining differently labeled samples in the final mixture.

9.5   Recognizing corresponding peptide variants

We have already mentioned that the quantification can be carried out at the MS level or on the MS/MS level. In both cases the basic quantification tasks are:

1. find the relevant peaks in the spectrum that are derived from different variants of the same peptide;

2. determine the intensities of these peaks.

Intensities of the peaks corresponding to variants of the same peptide are then compared to obtain a relative quantification. The first task varies with the approach used, and is the most complicated, while the second task is mainly carried out by applying standard procedures, see Chapter 4.4. In the following we briefly consider the first task, with further details covered in the succeeding chapters.

9.5.1   Recognizing peptide variants in MS spectra

To recognize peptide variants in MS spectra let the mass difference between the labels used for two variants be d. Then for a peptide with charge z and a number of labels l, the m/z difference between the variants will be dl/z.

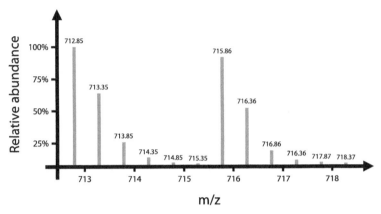

Figure 9.2   Illustration of an MS spectrum with SILAC labeling. The mass difference between the labels is 6 Da, the charge of the peptide is 2, and there is generally one label per peptide.

For given values of l and z one then searches for pairs of peaks that are this distance apart on the m/z scale, and one investigates if the two peaks can be variants of the same peptide. This investigation can, for example, include the comparison of the isotope patterns, or a comparison of the identifications obtained from the MS/MS spectra for which each of these peaks served as precursor mass. Figure 9.2 shows an example MS spectrum with peak pairs.

9.5.2   Recognizing peptide variants in MS/MS spectra

Labels for MS/MS quantification are all chemically incorporated. Most of the labels are *isobaric*, meaning that they have the same mass at the MS level. Because of their identical chromatographic properties, the different variants of a peptide occur at (approximately) the same position of the gradient and thus in the same MS spectrum. The peptide variants are then distinguished in the following MS/MS spectrum, where the fragmentation process causes the different labels to produce different patterns. Isobaric labels thus allow the simultaneous comparison of several samples in a single quantification run, a process referred to as multiplexing.

Isobaric labels allow different peptide variants to be recognized in the MS/MS spectra in one of two distinct ways, depending on the type of label used.

Use of reporter ions

This technique for differentiating peptide variants in an MS/MS spectrum is based on the dissociation of the labels in the fragmentation process, with each label producing a so-called reporter ion showing up as a separate peak at a fixed m/z-position. The reporter peaks are then used for quantification, and multiplexing is usually employed since the different reporter ions are chosen to be clearly distinguishable based on their unique m/z in the MS/MS spectrum. Even though they rely on quantification by proxy through the reporter ions, these methods are far more popular than fragment ion based MS/MS quantification methods. The use of reporter ions is described in detail in Chapter 10.

Use of fragment ions

In this technique the peptides are labeled at each terminus, and the labels stay connected to the fragments after fragmentation. The sum of the masses of the two labels at the termini are equal for each peptide variant (thus isobaric), but the labels at each terminus are different. In this way the fragments of the two peptide variants occur at different m/z-positions with a fixed distance. The fragment ions themselves are used for the quantification in this approach, but only two samples can be compared. The use of fragments ions for quantification is described in detail in Chapter 11.

9.6   Reference free vs. reference based

Label based quantification can be performed with or without a reference sample, where the reference sample is used for experimental normalization.

9.6.1   Reference free quantification

When two or more samples are compared in an LC-MS/MS run without a reference all the samples are called *experimental samples*, that is, samples to be analyzed. For a protein P all the pair-wise ratios $\frac{\text{abundance}(P_i)}{\text{abundance}(P_j)}$ are determined, where abundance(P_i) is the abundance of protein P in sample i.

The reference free setup has the advantage of using all sample slots for experimental samples, and therefore does not have to reserve a precious slot for a reference. But at the same time, the lack of a common reference is often a major drawback when samples need to be compared, as the data analysis becomes more complicated.

9.6.2   Reference based quantification

One of the samples in the LC-MS/MS run is used as a *reference r*, and the ratios $R_i = \frac{\text{abundance}(P_i)}{\text{abundance}(P_r)}$ are calculated for each of the experimental samples. This strategy is commonly used for reporter based MS/MS quantification, and can also be used in SILAC (see Chapter 12.1.1) and in SRM (see Chapter 15).

When used in SILAC or SRM the reference is often called an *internal standard*. The incorporation of such an internal standard into the samples is often referred to as a *spike-in*. The reference sample can be a protein sample (as in reporter-based and SILAC quantification), or a peptide sample (as in SRM).

Note that the pair-wise protein ratios are calculated via the reference ratios (ratio of ratios), in contrast to the reference free approach where they are directly calculated. The indirect calculation has a tendency to increase the quantification uncertainty.

However, the reference ratios are usually log-normally distributed, and a reference sample is therefore most often used.

Whether the pair-wise ratios $\frac{R_i}{R_j}$, or the reference ratios $\frac{\text{abundance}(P_i)}{\text{abundance}(P_r)}$ should be used in the downstream analysis depends on the statistics that will be used. If a T-test is used $\frac{\text{abundance}(P_i)}{\text{abundance}(P_r)}$ can be returned, however if a G-test is used $\frac{R_i}{R_j}$ should be calculated.

9.7   Labeling considerations

There are differences between the labeling techniques yielding method specific advantages and drawbacks. Some issues related to the choice of labels are:

- It is usually an advantage if the labels are incorporated as early as possible in the experimental procedure, as this reduces variability.

- MS/MS-based labeling allows for a higher degree of multiplexing as the detection of the peaks used for quantification is easier for MS/MS-based labeling than for MS-based labeling, especially when reporter ions are used.

- Many mass spectrometers have a higher sensitivity and accuracy at the MS level than at the MS/MS level. The reason is that the acquisition of MS/MS spectra often results in a limited amount of fragments, thus reducing the actual ion count. This points in favor of MS labeling.

- Some MS/MS spectra will contain high intensity fragments, while others will only contain low intensity fragments. This strongly influences the accuracy of the measured ratios, as strong signals give more reliable ratios.

- MS level quantification tends to have higher precision than MS/MS level quantification, so it becomes easier to detect differences in protein expression (higher power).

9.8   Exercises

1. Assume that we have an MS spectrum from a label based quantification experiment where the following peaks have been found in the m/z-interval [546, 556]:

   a:547.61, b:548.13, c:550.12, d:550.61, e:551.11, f:553.13, g:554.62, h:555.13.

   Also assume that the peptides can have a charge of two or three, and carries one or two labels. The mass of the labels are 6 Da.
   (a) Find the pairs of peaks that can be from a (labeled, unlabeled) pair of the same peptide, and show the charge and the number of labels. Take into account that the distance should not deviate from the theoretical distance by more than 0.02 m/z units.

(b) Use isotope envelopes to determine which of the pairs can be real pairs, requiring at least two succeeding isotopes. Determine the masses of the peptides represented by the pairs.

(c) Describe how the intensities of the peaks could be used to further verify the results found in (b).

2. We have a label based, reference free quantification experiment. We manage to determine numerous (labeled, unlabeled) peptide pairs, and calculate the abundances. Which type of statistical analysis would you perform to discover differentially abundant peptides?

9.9   Bibliographic notes

Labeling techniques   Ong and Mann (2005), Bantscheff et al. (2007b), Zhang and Neubert (2006).

10

Reporter based MS/MS quantification

Reporter based MS/MS quantification is a multiplexed quantification approach where small parts of the labels, referred to as reporters, are cleaved off during the fragmentation process. They then occur as separate peaks at fixed m/z positions in the MS/MS spectra, and these peaks are used as a proxy to quantify the corresponding peptide variants. In order to ensure easy detection of the reporters they should occur in areas of the spectrum where there are few or (preferably) no other peaks. This reduces the chance of the reporter ions interfering with the rest of the spectrum (or vice versa), allowing the reporters to be quantified accurately, and the spectrum to be identified as usual. Figure 10.1 shows an example spectrum with iTRAQ reporter peaks.

10.1 Isobaric labels

In order to construct labels that have the same mass at the MS level (isobaric labels), but different masses at the MS/MS level, the (complete) labels consist of three groups:

- *Reporter group*: With a different mass for each label.

- *Balance group*: Also with a different mass for each label, but such that the combined mass of the reporter and balance groups is identical for all labels.

- *Reactive group*: Required to bind the labels to the peptides. This part is cleaved off during the labeling of the peptide.

Figure 10.2 schematically illustrates a label containing the three groups described above.

Computational and Statistical Methods for Protein Quantification by Mass Spectrometry,
First Edition. Ingvar Eidhammer, Harald Barsnes, Geir Egil Eide and Lennart Martens.
© 2013 John Wiley & Sons, Ltd. Published 2013 by John Wiley & Sons, Ltd.

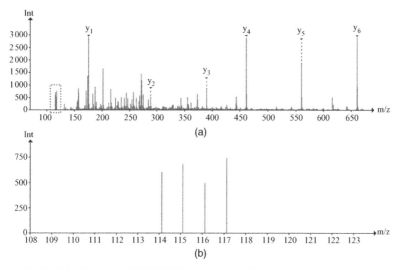

Figure 10.1 (a) shows a full MS/MS spectrum with the y-ions annotated, and the region for the reporter peaks indicated. (b) zooms in on the reporter region, highlighting the reporter peaks at positions 114.1 to 117.1. For this illustration only the reporter peaks are shown.

The combination of a unique reporter group and a specifically matched balance group ensures that different variants of the same peptide occur at the same position in the MS spectra. The balancer group is therefore sometimes referred to as the *normalization group*.

We will first describe the two most commonly used labels, iTRAQ and TMT, and then describe how quantification using such labels can be performed. Note that

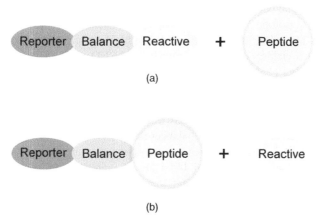

Figure 10.2 Illustration of the different groups of a reporter label. (a) before binding to the peptide, (b) after binding, with the peptide taking the place of the reactive group.

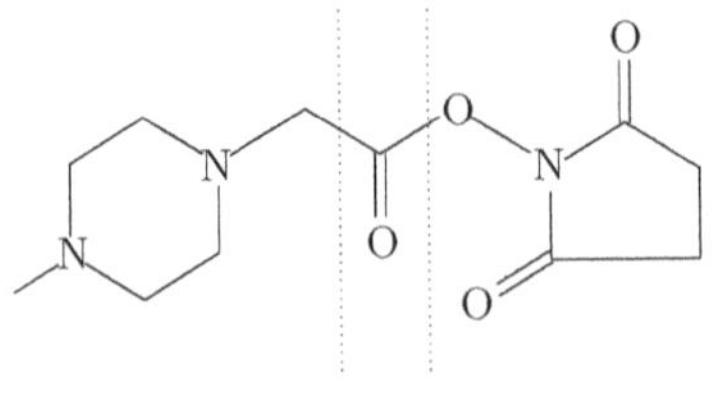

Reporter   Balance   Reactive

Figure 10.3   Illustration of the iTRAQ™ 4-plex label. Based on Pichler et al. (2010).

some mass spectrometers (especially ion traps) have problems recording peaks at low m/z-values, where the reporter ions of iTRAQ and TMT are located.

10.2   iTRAQ

The currently most common reporter based labels are iTRAQ and TMT, briefly introduced in Chapter 9.5.2. iTRAQ can be used to simultaneously compare four or eight samples, while TMT supports six parallel samples. In Figure 10.1, an example iTRAQ MS/MS spectrum was shown.

The iTRAQ labels are attached to the peptides by binding to free amines; either the N-terminal amino group of the peptide, or an ϵ amino group of lysine. During binding the reactive group of the label is cleaved off. The number of labels that can be attached to a peptide is typically equal to one for the N-terminus of the peptide, plus one additional label for each lysine. For tryptic peptides without missed cleavages, this number is zero or one.

Figure 10.3 shows the chemical structure of an iTRAQ 4-plex label, where 4-plex refers to the fact that these labels can be used to compare up to four samples simultaneously. The naturally occurring monoisotopic chemical formula (thus using the ^{12}C and ^{14}N isotopes) for the label (reporter + balance groups) is $C_7N_2OH_{13}$, amounting to 141.1 Da, and the chemical formula for the reporter group alone is $C_6N_2H_{13}$, which yields a monoisotopic mass of 113.1 Da.

When the label is attached to a free amine of a peptide, one hydrogen atom is removed from the peptide, thus yielding a net increase in the peptide mass of 140.1 Da if the naturally occurring monoisotopic label was attached. However, since different versions of the reporter group should have different masses, heavy stable isotopes are used to create different reporter and balancer masses.

In 4-plex iTRAQ, heavy isotopes are incorporated to create a mass increase of 4 Da, yielding a total label mass of 145.1 Da. By spreading the isotopes over the reporter and balancer groups in different ways, different masses can be obtained for the reporters, and complementary masses for the balancers, while maintaining the same total mass. Once incorporated into a peptide, the net mass increase amounts to 144.1 Da for each attached label.

Table 10.1   Example of distribution of heavy isotopic masses between reporter and balancer groups. The total additional mass for each label variant is always 4 Da.

Label	Reporter	Balance
114	^{13}C	$^{13}C^{18}O$
115	$^{13}C\ ^{15}N$	^{18}O
116	$^{13}C_2\ ^{15}N$	^{13}C
117	$^{13}C_3\ ^{15}N$	

Table 10.1 shows an example of how the different heavy isotopes can be distributed to obtain reporter group masses of 114.1, 115.1, 116.1, and 117.1 Da, along with the corresponding balancer groups.

The formula for the 115-label is $C_5\ ^{13}C_2N_2\ ^{18}OH_{13}$ (145.1 Da). The corresponding reporter group has formula $C_4\ ^{13}C_2N\ ^{15}NH_{12}$, yielding a mass of 115.1 Da.

The mass of the reporter ions can be more accurately calculated to 114.111, 115.108, 116.112, 117.115 Da.

10.2.1   Fragmentation

The iTRAQ-labels were originally designed primarily for use with collision induced dissociation (CID) fragmentation. Standard CID fragmentation of iTRAQ labels is well suited to commonly used instruments such as a Q-TOF or TOF/TOF.

Fragmentation breaks the labels at the bond between the reporter group and the balancer group (see Figure 10.3), but also at the bond where the balancer group is attached to the peptide. The reporter group retains the charge in this fragmentation, thus producing ions called *reporter ions*. They are also sometimes referred to as *marker ions* or *signature ions*. If fragmentation also occurs at the bond between the balancer group and the peptide, a neutral loss of the balance group occurs.

As a result of these additional fragmentation sites in the label, new types of fragment ions can occur in the spectra that are composed of standard fragment ions with a portion or all of the label still attached. These additional ions ought to be accounted for as well, and thus render the identification process a bit more complicated than usual.

Hybrid fragmentation

In CID mode ion trap instruments are unable to see ions at low m/z values. This limitation also holds for hybrid instruments where ion traps are used for fragmentation, such as the LTQ-FT and LTQ-Orbitrap mass spectrometers.

The LTQ-Orbitrap in particular has recently become very popular, and this instrument is therefore schematically illustrated in Figure 10.4. One popular configuration

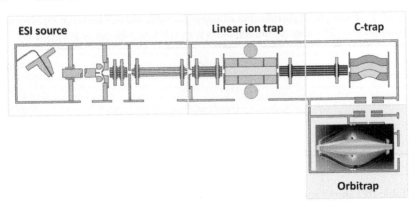

Figure 10.4   Schematic illustration of an LTQ-Orbitrap. Reprinted from Scigelova and Makarov (2006) by permission of John Wiley & Sons.

of this spectrometer uses the Orbitrap for the MS level, and the ion trap for the MS/MS level, providing high mass accuracy and precision for the intact peptide in the MS scan, but much less accuracy and precision for the fragments in the MS/MS scans.

The problem with CID fragmentation in ion traps has resulted in other fragmentation techniques that allow for the analysis of the low m/z region used in parallel with CID, especially *HCD (Higher energy C-trap Dissociation)* and *PQD (Pulsed Q Dissociation)*. Both produce essentially the same fragment ion types as CID, yet allow the low mass range to be recorded in hybrid instruments.

For this approach two versions of each MS/MS spectrum are produced; one using HCD or PQD producing spectra appropriate for quantification, and one using CID producing spectra appropriate for identification. These two spectra can then be used independently, both in the quantification and the search program used. Software performing both operations simultaneously also exists, and the two spectra can then either be combined into one hybrid spectrum before processing, or the combination can be handled internally by the software.

Because the identification scores derived by the search programs often depend on the relative intensities of the identified peaks in the spectrum, the reporter ions (that are designed to be distinct from any sequence-specific ions) can influence the relative intensities. Indeed, highly intense reporter ions would be very useful for quantification, but would provide large, unassigned peaks in the identification stage, thus lowering the score there. Reporter ion intensities are therefore typically normalized to one in the combined spectra.

Finally we will also mention *ETD (Electron Transfer Dissociation)* which results in different fragmentation sites compared to CID, producing mostly $c-$ and $z-$ fragment ions. It has been shown that ETD is compatible with iTRAQ, which can be valuable due to its suitability for the identification of peptides with certain post-translational modifications.

10.2.2   Reporter ion intensities

The basic quantification operation is based on the identification of reporter ion peaks in the appropriate m/z region, and the subsequent calculation of their intensities. This can be performed in the following way by standard techniques:

1. Search for peak apexes in the regions $rm \pm k$ where rm is the m/z of the reporter ions (e.g., 114.1 etc.) and k is a constant depending on the accuracy of the mass spectrometer (e.g., 0.2).

2. Subtract background noise.

3. Calculate the intensity of each peak, as discussed in Chapter 4.4.

Isotope interference

Several isotopes of a reporter ion may occur, for example, isotopes of the 114-reporter can occur at nominal m/z-values 113, 114, 115, 116. These different isotopic variants occur for each label because it is extremely difficult to create chemicals that contain only one type of isotope. The observed abundances should therefore be corrected accordingly. Let:

- x_i be the total intensity of reporter ion i;

- $e_{i,j}$ be the fraction of x_i that occurs at the m/z position for reporter ion $i + j$ (the isotope distribution), $j = -2, \ldots, 2$;

- y_i be the observed intensity at the m/z position for reporter ion i.

We then see that

$$e_{i,0} = 1 - \sum_{j \in [-2,-1,1,2]} e_{i,j}. \tag{10.1}$$

We assume that y_i is only constituted by the reporter ions. In practice other peaks can also contribute interfering intensity, but it is quite difficult to estimate the amount of this almost random interference in a particular spectrum.

Then we have

$$y_i = \sum_{j=-2}^{2} x_{i+j} e_{i+j,i}; \; i = 114, \ldots, 117, \tag{10.2}$$

for 4-plex iTRAQ.

The values of the $e_{i,j}$'s are called *correction factors*. The correction factors provided by Applied Biosystems (the producer of the iTRAQ labels) as of July 2009 are shown in Table 10.2.

From this we can formulate the linear equation system in Table 10.3, which can be solved for the x_i's.

Table 10.2   Correction factors for iTRAQ 4-plex as of July 2009.

i\j	−2	−1	1	2
114	0.0	0.01	0.059	0.002
115	0.0	0.02	0.056	0.001
116	0.0	0.03	0.045	0.001
117	0.001	0.04	0.035	0.001

Table 10.3   The linear equation system using the correction factors in Table 10.2.

$$y_{114} = 0.929x_{114} + 0.002x_{115} + 0.000x_{116} + 0.000x_{117}$$
$$y_{115} = 0.059x_{114} + 0.923x_{115} + 0.030x_{116} + 0.001x_{117}$$
$$y_{116} = 0.002x_{114} + 0.056x_{115} + 0.924x_{116} + 0.004x_{117}$$
$$y_{117} = 0.000x_{114} + 0.001x_{115} + 0.045x_{116} + 0.924x_{117}$$

Example If all the measured intensities (y_is) are 1, the corrected intensities (x_is) become (1.055, 0.984, 0.976, 1.034). This example is taken from Arntzen et al. (2011).
△

10.2.3   iTRAQ 8-plex

The exact chemical structure of the iTRAQ 8-plex label is not publicly available, but the 'basic' chemical formula for the label is $C_{14}N_4O_3H_{25}$, corresponding to 297.2 Da. The formula for the reporter group is the same as for iTRAQ 4-plex. The number of heavy isotopes used is eight. This means that the mass of the combination of the reporter and balance groups is 305.2 Da, and the mass of a labeled peptide is increased by 304.2 Da for each label.

Table 10.4 shows how the heavy isotopes can be distributed in order to achieve the desired reporter masses.

The mass of 120.1 Da is not used for a reporter ion in order to prevent interference by the phenylalanine immonium ion (at a mass of 120.08 Da).

Table 10.4   Reporter masses and isotopes for iTRAQ 8-plex.

Reporter masses	113.1	114.1	115.1	116.1
Reporter isotopes	-	^{13}C	$^{13}C^{15}N$	$^{13}C_2^{15}N$
Reporter masses	117.1	118.1	119.1	121.1
Reporter isotopes	$^{13}C_3^{15}N$	$^{13}C_3^{15}N_2$	$^{13}C_4^{15}N_2$	$^{13}C_6^{15}N_2$

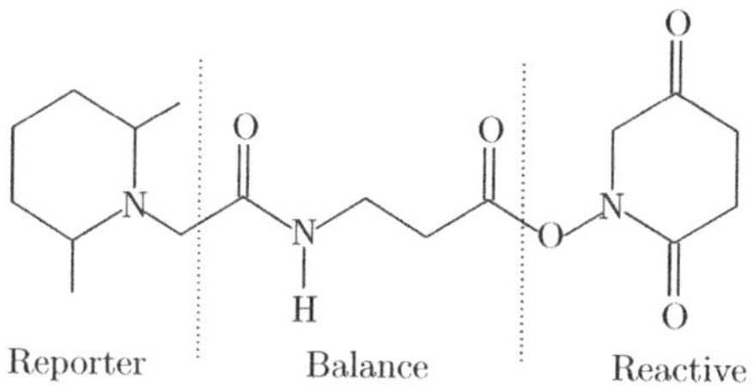

Figure 10.5   The structure of 6-plex TMT.

10.3   TMT – Tandem Mass Tag

TMT is an isobaric label that works very much like iTRAQ. The labels are bound to the free amines at the peptide amino-termini or at lysines, and during fragmentation reporter ions are produced. TMT exists as a 2-plex or a 6-plex set of labels. Figure 10.5 shows the structure of 6-plex TMT.

The naturally occurring monoisotopic formula of the label is $C_{12}N_2O_2H_{21}$, resulting in a mass of 225.2 Da, and the corresponding formula for the reporter is C_8NH_{16} which is 126.1 Da. Five heavy isotopes are used to get six different reporter ions. This results in a mass range for the reporter ions from 126.1 to 131.1 Da, a total label mass of 230.2 Da, and a net increase in peptide mass of 229.2 Da for each label attached.

Reporter ions for TMT 2-plex have a mass of 126.1 and 127.1 Da, and the increase in peptide mass for each label is 225.2 Da.

Programs that analyze spectra with TMT labels work in the same way as those aimed at analyzing iTRAQ labels.

10.4   Reporter based quantification runs

Figure 10.6 illustrates a reporter based quantification run with four samples.

The figure illustrates that in addition to the assumptions in Chapters 5.11 and 9.4 the main principle behind reporter based quantification is that *the relative intensities of the reporter ions will ideally mirror the relative abundances of the variants of the peptide under consideration*. In the example in the figure, the peptide variant with the highest abundance ought to be in sample two, as this sample generated the highest reporter ion peak (L2). This requirement in turn relies on an effective fragmentation process, which was discussed in more detail in Section 10.2.1.

10.5   Identification and quantification

For each peptide MS/MS spectrum there are two main tasks:

- Recognize the peaks corresponding to the reporter ions, and calculate their intensities.

Four–sample reporter–based quantification run

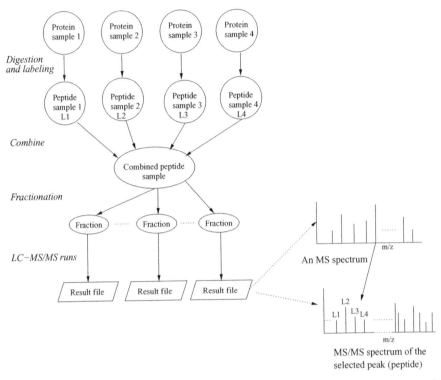

Figure 10.6   Illustration of a four-sample reporter based quantification run with fractionation, in which the four samples are labeled with the labels L1, L2, L3, L4 respectively. The reporter groups of these labels then occur in the MS/MS spectra as illustrated.

- Identify the peptide that generated the spectrum, using the other peaks in the spectrum.

These tasks can be performed separately and in any order, or in an interweaving manner. When determining the structure of a program for this task, the following should be considered:

- If a spectrum is not identified there is no reason to quantify it.

- If a spectrum is identified with low scoring/significance, there is probably little reason to quantify it.

- If a spectrum is not quantifiable, that is, no reporter peaks observed, the value of identifying it is questionable. One reason for still identifying it is that it may be used in the calculation of the confidence of the (protein) identifications.

Specifically for reporter based spectra we can also add:

- There can be spectra with low signal to noise ratios for the reporter ions.

- There can be spectra where some of the reporter ions have reached the maximum intensity threshold of the detector used, causing these observed intensities to deviate from the corresponding peptide variant abundances.

10.6   Peptide table

After identification and quantification of the MS/MS spectra, the results can be compiled into a peptide table, as explained in Chapter 5.10, with one row for each MS/MS spectrum that can take part in the calculation of protein ratios.

Each row in the table should contain:

- the identified peptide(s), remember that an MS/MS spectrum can have a significant score against more than one peptide;

- score(s) and significance(s) of the identification(s);

- the identified protein(s);

- whether the identified peptide is unique to one protein, or included in several proteins;

- the intensities of each of the peptide variants (as derived from the peaks of the reporter ions).

10.7   Reporter based quantification experiments

A general reporter based quantification experiment consists of analyzing n situations, but most often $n = 2$, which is what we will consider here. For each situation several biological and technical replicate samples should be included. Given that the number of samples that can be analyzed in a single quantification run is limited (at the time of writing, maximum eight), several quantification runs have to be performed when the number of samples is larger. This is illustrated in Figure 10.7.

For each quantification run an LC-MS/MS run is performed for each fraction, producing a peptide table. These tables are used in the statistical analysis to present differentially abundant proteins.

10.7.1   Normalization across LC-MS/MS runs – use of a reference sample

A common way of normalizing data from different sources is to keep one sample constant in each source, and then normalize the data relative to the data from this sample. In our context this means that there should be a common protein sample included in each run, and the relative abundances mentioned above should be calculated relative

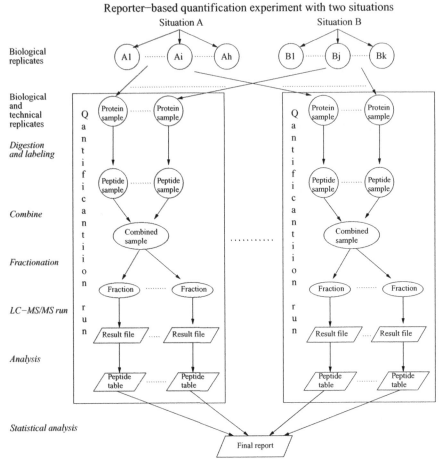

Figure 10.7 Illustration of a quantification experiment for two situations using several quantification runs.

to this sample, called the *reference sample*. To avoid variation due to the different labels the reference sample should be labeled by the same label in each run.

The reporter of this label is called the *reference reporter*. For each spectrum the intensities of the other reporter ions are therefore divided by the intensity of the reference reporter to achieve *reference relative intensities*.

A reference sample can be chosen as one of the samples or as a pooled sample, typically created by combining all the original samples. The pooled sample should ensure that the reference sample contains all proteins, and that the reference reporter peak can always be found in each spectrum where at least one other reporter ion is observed. However, one undesired consequence of this procedure is that rare proteins could become so diluted that their quantification becomes unreliable. Nevertheless, an investigation by Song et al. (2008) showed no negative impact of using a pooled sample as the reference sample. It is crucial for the result that the reference sample is exactly the same in all runs.

It has been observed that the reference relative intensities follow a log-normal distribution, and a log transformation is therefore usually performed in order to obtain a normal distribution. As explained in Chapter 6.6, the $\log_2$ transformation is most often used for this purpose.

Is a reference sample necessary?

The use of a reference sample reduces the number of samples that can be analyzed in a single quantification run by one. It is therefore reasonable to ask if a reference sample is really necessary. If a reference sample is not used, each spectrum has to be separately normalized as explained in Chapter 7.2, thus information from the whole spectrum is used. Normalization by a reference sample on the other hand implies that only the considered peaks (two in each normalization operation) are used, rendering the normalization simpler and more reliable.

10.7.2   Normalizing within an LC-MS/MS run

Normalization within an LC-MS/MS run has to be performed if the observed intensities may be biased by one or more of the reporters. As previously pointed out most of the protein abundances are unchanged across the samples. Two methods utilizing this assumption for normalization are described below.

Use of total intensity

The intensities are summed over all reporter ions and MS/MS spectra (or alternatively, only over the identified spectra) for each sample, and these sums are normalized to become equal.

Use of relative peak intensity distributions

For each sample the distribution of the log-transformed reference relative peak intensities are determined. Since most of these should ideally be zero, normalization is performed to move either the mean or the median of each distribution to zero. The median is typically preferred since it is more robust to outliers.

This normalization also has the effect that the distributions become symmetric around zero. Another option is to normalize the distributions such that all quartiles become equal. This is used in Keshamouni et al. (2007), where a monotone piece-wise linear function is used for this purpose.

10.7.3   From reporter intensities to protein abundances

Figure 10.8 illustrates that abundance information for a given protein in a quantification run can occur in numerous MS/MS spectra, from multiple LC-MS/MS runs (in the figure h runs).

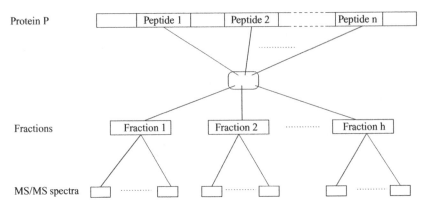

Figure 10.8   The abundance of a given protein in a quantification run is distributed across many MS/MS spectra.

The path from reporter intensities in individual spectra to protein abundances can be performed in several ways. Let

- $a_i, i = 1, \ldots, n$ be the intensities of a (normalized) reporter ion corresponding to the protein and sample under consideration;

- e_i be the intensity of the reference reporter corresponding to a_i;

- $r_i = \frac{a_i}{e_i}$ be the reference relative reporter abundance;

- R is the log of the reference relative protein abundance.

One alternative is to calculate relative peptide abundances for all the identified peptides of the protein, and then the protein abundance from these peptide abundances. However, due to the normalization performed one should ideally calculate the protein abundance directly from the log relative reporter abundances, utilizing one of the formulas presented in Chapter 5.8.

Correlation is the most commonly used approach for this purpose, with the a_i's on the vertical axis, and the e_i's on the horizontal axis. The slope s of the correlation line is calculated, and $R = \log_2 s$. We know from Figure 5.8 in Chapter 5.8 that the variance is largest for small intensities, thus variance stabilization can be performed before correlation, or spectra with low reporter intensities can be discarded altogether.

Weights depending on the confidence of the identification have also been used, both for weighted average and weighted correlation calculations.

10.7.4   Finding differentially abundant proteins

From Figure 10.7 it should now be clear that we have calculated the reference relative abundances across a set of replicates for each identified protein P. Let A_i

be the abundance of P for replicate i of situation 1, and B_i the abundance for replicate i of situation 2. Different statistics can be used to analyze the abundance of P in the two situations, as described in Chapter 8.1.

We can calculate the averages of the A_is and the B_is, $\bar{A}$ and $\bar{B}$. These are normally distributed due to the assumption of log-normal distribution. We calculate $\bar{A} - \bar{B}$ for each protein, and we note that the difference of two normal distributions is again normally distributed. To estimate the significance of P being differently abundant we can then use the Z-variable. Since we have replicates for each protein, we can also use a T-statistic to compare the differences of the two mean values. Do note that, due to the large number of proteins that are typically tested for such experiments, multiple testing correction must always be employed, see Chapter 8.5.

10.7.5 Distributing the replicates on the quantification runs

As described in Chapter 5.2 replicates are often required in quantification experiments. The replicates and the reference sample should be distributed across the runs and *channels* (the term 'channel' is often used to refer to a specific label) of the labels to optimize the handling of the technical and biological variabilities. Some simple rules used for the distribution of the samples across the channels are:

- the reference sample should be fixed to one specific channel;
- the replicates should be spread over the channels, ensuring that no sample is ever run on the same channel.

When replicates are spread across channels, the replicate tests both the technical reproducibility of the replicate, as well as the reproducibility of quantification by the different reporters; so if there is a difference, we know what it is, not where it comes from.

Example Suppose we have two situations A and B, and three biological replicates for both of them, A1, A2, A3 and B1, B2, B3, and a reference sample R.

We use two technical replicates for each biological replicate, resulting in a total of 12 replicates, spread over four iTRAQ runs. We then distribute the replicates as shown in Table 10.5.
△

From the example we see that:

- the technical replicates occur in different channels in each run;
- the number of pairs of biological replicates that occur together in more than one run is minimized.

However, there will almost always be some co-occurrences, (A2, B2) occurs together in two runs, but this double co-occurrence could be put to use. These ratios

Table 10.5   Example of the distribution of replicates across iTRAQ runs.

iTRAQ run	114	115	116	117
1	A1	A2	B1	R
2	B2	B3	A3	R
3	A2	B1	B2	R
4	B3	A3	A1	R

should always be the same, and if they are not they could be used to estimate the error.

10.7.6   Protocols

We know that the result of quantification experiments to a high degree depends on the quality of all the individual steps in the process, and thus on having access to high quality protocols to follow. One such protocol is described in Unwin et al. (2010).

10.8   Exercises

1. Explain the equation for y_{115} in Table 10.3.

2. We consider two quantification runs, and for simplicity we only look at one spectrum from each, s1 and s2. We get intensity values for three peaks in each spectrum:

	s1	s2
Highest intensity peak	9200	5400
Reference peak	6300	4600
Sample peak	7200	2100

Calculate the ratio between the sample peaks (i) when normalization via the reference is used, and (ii) when spectral normalization using the highest peak is used. Which result would you trust the most?

3. Explain how the reporter ions can negatively affect the calculation of the (spectrum) identification score.

4. In Section 10.5 it is stated that it might be useful to identify a spectrum even though no reporter peaks are observed. Explain why.

5. We have identified four peptides from a protein P, and have the following abundances:

Peptide	Reference reporters	Sample reporters
1	7.2, 6.8, 9.2	14.1, 13.9, 18.6
2	6.4, 5.2, 4.6, 8.2	12.5, 6.1, 9.1, 16.5
3	9.2, 5.8, 4.6	10.1, 6.3, 4.9
4	8.4, 6.2, 5.7	17.1, 12.5, 11.2

As an example this means that peptide 1 is identified by three MS/MS spectra, and in the first spectrum the intensity of the reference reporter is 7.2, and the reporter intensity corresponding to the sample under consideration is 14.1.

(a) Determine if any of the peaks in any of the spectra seem to be outliers, and also if any of the peptides can be considered outliers.

(b) Remove the suspected outliers, and calculate the log of the reference relative protein abundance using any of the procedures in Chapter 5.8.

10.9   Bibliographic notes

Software   There is a lot of software for analyzing reporter based data, especially for iTRAQ data, some are listed below.

- Mascot (http://www.matrixscience.com).
- SpectrumMill (http://spectrummill.mit.edu).
- ProteinPilot (http://www.absciex.com/Products/Software/ProteinPilot-Software).
- Multi-Q (http://ms.iis.sinica.edu.tw/Multi-Q-Web).
- VEMS (http://www.portugene.com/vems.html).
- Scaffold   (http://www.proteomesoftware.com/Proteome_software_pub_with_scaffold.html).
- MassTRAQ (http://arnetminer.org/viewpub.do?pid=121019).
- IsobariQ (http://folk.uio.no/magnusar/isobariq).
- PROTEOIQ (http://www.bioinquire.com).
- jTraqX (http://sourceforge.net/projects/protms/files/jTraqX).

Statistics   More advanced methods for statistical calculations related to iTRAQ can be found in Hill et al. (2008), Oberg et al. (2008).

iTRAQ   Ernoult et al. (2008), Pichler et al. (2010), Boehm et al. (2007), Ross et al. (2004), Phanstiel et al. (2008), Koehler et al. (2009), Bantscheff et al. (2007a), Gan et al. (2007), Ernoult et al. (2008), Lacerda et al. (2008), Unwin et al. (2010) Zhang and Neubert (2006), Keshamouni et al. (2007), Lin et al. (2006), Karp et al. (2010), Muth et al. (2010).

TMT   Thompson et al. (2003), Dayon et al. (2008), Pichler et al. (2010), Vaudel et al. (2010).

Experimental design   Song et al. (2008).

Fragmentation   Pichler et al. (2010), Phanstiel et al. (2008), Dayon et al. (2008), Armenta et al. (2009).

Normalization   Keshamouni et al. (2009).

Technical considerations   Zhang and Neubert (2006).

11

Fragment based MS/MS quantification

In fragment based MS/MS quantification the peptides are labeled at each terminus, and the labels stay connected to the fragments after fragmentation. The sum of the masses of the two labels at the termini are fixed and equal for each peptide variant (and the total modification is thus isobaric), but the labels at each terminus are chosen to be different for each variant. Corresponding fragments of peptide variants thus occur at different m/z positions, separated by a fixed distance.

The recognition of corresponding fragments thus resembles the recognition of peptide variants in labeled MS spectra. But since there is only one label per fragment, and in many cases only one charge, the recognition is easier. The fragment ions themselves are then used for quantification.

Use of the fragment ion based strategy overcomes two main drawbacks of reporter ions:

- the incompatibility with ion trap instruments due to the low mass of the reporter ions;

- despite efforts to design the reporter ion mass so that interference with other (nonreporter) fragments is minimal, interference can occur, negatively influencing the quantification accuracy and precision.

11.1 The label masses

In the remainder of this chapter we consider the comparison of two samples. And though it is theoretically possible to perform multiplexing, the complexity of the

Computational and Statistical Methods for Protein Quantification by Mass Spectrometry, First Edition. Ingvar Eidhammer, Harald Barsnes, Geir Egil Eide and Lennart Martens. © 2013 John Wiley & Sons, Ltd. Published 2013 by John Wiley & Sons, Ltd.

MS/MS spectra would increase to such an extent that a correct interpretation would become very difficult.

- Let $m_{1,N}$, $m_{1,C}$ be the masses of the labels attached to the N-terminus and the C-terminus of the peptide versions in sample 1, and $m_{2,N}$, $m_{2,C}$ the corresponding masses for the labels in sample 2.

- These masses are chosen such that $m_{1,N} + m_{1,C} = m_{2,N} + m_{2,C} = M\ Da$ (the total modification is isobaric at the MS level).

- $m_{2,N} - m_{1,N} = m_{1,C} - m_{2,C} = k\ Da$, meaning that the pairs of b-ions and pairs of y-ions appear in the MS/MS spectra separated by k m/z-units when assuming a single charge. More generally, the corresponding fragment ions will be $\frac{k}{c}$ m/z-units apart for a charge of c. Note however that the variants occur in a different order in each ion type; if the ions from sample 1 appear at the highest m/z position for the y-ions, then the ions from sample 2 will appear at the highest m/z position for the b-ions.

Figure 11.1 shows an example MS/MS spectrum. b_3-ions are identified at $m/z = 372$. A smaller peak occurs just to the right, which can be from the other variant. For the y-ions the smaller peaks occur to the left of the larger peaks.

IPTL (Isobaric Peptide Termini Labeling) is one fragment-based labeling technique that can be used to compare two samples. Figure 11.2 shows the four labels that are used in this approach.

The masses of the labels are as follows:

- MDHI: $(C_4N_2OH_8)$ 100 Da;

- MDHI-d4: $(C_4N_2OH_4D_4)$ 104 Da;

- SA: $(C_4O_3H_4)$ 100 Da;

- SA-d4: $(C_4O_3D_4)$ 104 Da.

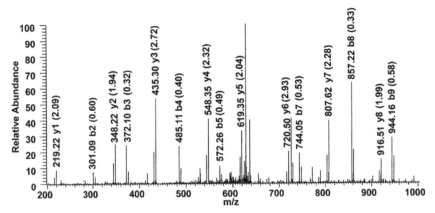

Figure 11.1 An MS/MS spectrum after use of IPTL. Reproduced from Koehler et al.
© 2009 with permission from American Chemical Society.

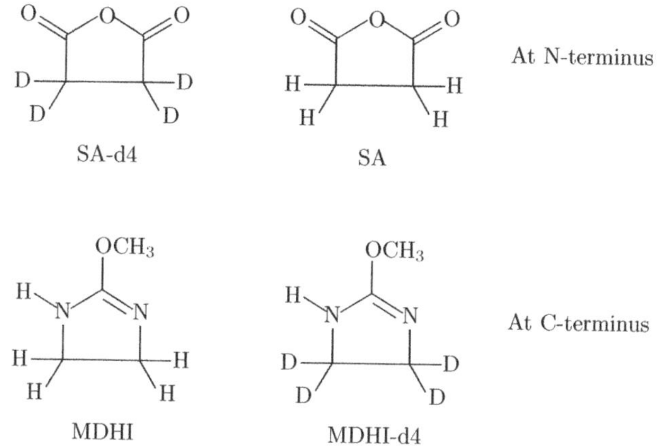

Figure 11.2   The four labels used in IPTL. Reproduced from Koehler et al. © 2009 with permission from American Chemical Society.

The peptides are labeled using the following combinations of labels:

- Sample 1: SA at the N-terminus and MDHI-d4 at the C-terminus;

- Sample 2: SA-d4 at the N-terminus and MDHI at the C-terminus.

During the labeling reaction CH_4O is lost from MDHI and MDHI-d4, reducing the mass of these labels with 32 Da. SA and SA-d4 are tagged as complete molecules. The attached labels thus yield the following masses:

- $m_{1,N} = 100\,Da$, $m_{2,N} = 104\,Da$, $m_{1,C} = 72\,Da$, $m_{2,C} = 68\,Da$;

- $k = 4\,Da$, $M = 172\,Da$.

The labeling requires lysines at the C-terminus, and Lys-C is therefore used for digestion instead of trypsin. However, a novel derivative of the method exists, using trypsin as the protease and C-terminal carboxyl-based labeling rather than lysine-labeling.

Fragment based quantification implies that the quantification is performed using the same peaks as the identification. The number of peaks is theoretically doubled compared to the unlabeled spectra, which makes the identification of the spectra more complex. On the other hand, the pairs of peaks add more information and can also be used to increase the confidence in the identification, since all sequence specific ions should be found in pairs a known distance apart.

11.2   Identification

General search programs (such as Mascot) can be used for spectrum identification. The search should be performed with the masses of the labels as variable modifications.

Although the labeling increases the number of peaks and thus makes the spectrum more complicated to interpret, there are also clear benefits to this approach. The different behavior of the *b* and *y*-ions can be used to distinguish these ions types, making identification simpler or more reliable.

To allow this distinction, all pairs of peaks that are k m/z units apart (assuming a single charge) are found. If the intensities of the two peaks in all the pairs are largely similar, the peptide probably belongs to a protein that is not differentially abundant, and the distinction between the ion types cannot easily be made. If the intensities are however different, we can divide the pairs into two classes: The first where the peak with lowest intensity has the lowest m/z, and the second where the least intense peak is found at the highest m/z. These two classes should then represent the different ion-types. Do note, however, that this method relies on the assumption that all fragment ions will correctly reflect the relative intensity of the parent peptide variant.

11.3   Peptide and protein quantification

In contrast to reporter based quantification, where peptide variant quantification is determined by a single peak for each version, the fragment based quantification is derived from a number of fragment intensities. Each recognized *b*- and *y*-ion in an identified spectrum can be used in the calculation.

It is reasonable to calculate the peptide variant abundance ratio, and the formulas in Chapter 5.8 can be used for this purpose. In the program *IsobariQ* Arntzen et al. (2011), the logs of the fragment ion intensity ratios are used, and the median of these log-transformed ratios is taken as the peptide (spectrum) ratio. We note that the ratio is thus calculated from the original measured values at the fragment ion level, arguably the lowest level of information in the spectrum. Contrast this with programs for iTRAQ quantification, where the ratio is directly calculated at the higher peptide or protein level.

From the peptide ratios the protein ratios can be calculated using the formulae and methods explained in 5.8, while always taking care to correctly handle outliers.

Statistical significance for a potentially differentially expressed protein can then typically be calculated using the methods described in Chapter 8.2, as always taking care to correct for multiple testing.

11.4   Exercises

1. Consider the spectrum in Figure 11.1.

   (a) Try to identify corresponding peaks from the two variants, and determine the intensity values.

   (b) Calculate the intensity ratios.

   (c) Are there any likely outliers?

(d) Compare the ratios of the y-ions to the ratios of the b-ions regarding mean and variance. Perform tests (if necessary) to see if they are equal.

(e) Calculate a value for the peptide ratio.

(f) Try to determine the sequence of the peptide.

11.5   Bibliographic notes

IPTL   Koehler et al. (2009), Arntzen et al. (2011).

12

Label based quantification by MS spectra

Label based quantification by MS spectra consists of two main steps: Quantification by MS spectra and identification by MS/MS spectra.

The main principle for the quantification is straightforward: (i) the samples to be compared are modified using labels that have the same chromatographic characteristics, thus ensuring that the different versions of a peptide occur in the same MS spectra; (ii) the masses of the labels and the charge of the peptide are then used to locate variants of the same peptide in an MS spectrum.

12.1 Different labeling techniques

Various labeling techniques were listed in Table 9.1, and these will be described in more detail here. Note that there is a strong connection between the protease used for digestion and the labeling technique employed in order to achieve optimal peptide labeling.

12.1.1 Metabolic labeling – SILAC

SILAC (Stable Isotope Labeling with Amino acids in Cell culture) is a metabolic labeling technique. The labeling of the proteins is performed by providing cells with particular amino acids (one or two) carrying different isotopes. Commonly one of the amino acids consists of natural isotopes, whereas the others contain heavier stable isotopes.

Computational and Statistical Methods for Protein Quantification by Mass Spectrometry,
First Edition. Ingvar Eidhammer, Harald Barsnes, Geir Egil Eide and Lennart Martens.
© 2013 John Wiley & Sons, Ltd. Published 2013 by John Wiley & Sons, Ltd.

When trypsin is to be used for digestion the labeled amino acids are typically lysine and arginine, thus achieving a single label for most of the peptides. It should be noted that care is usually taken to use essential amino acids that are not readily metabolized or converted into other amino acids by the cell. If the labeled amino acid is converted into other amino acids, these too will be incorporated into future protein chains, which makes it much harder to predict the number of labels carried by a particular peptide.

For double SILAC (comparing two samples, often just referred to as SILAC) a common labeling is $^{13}C_6$ $^{15}N_2$ for lysine, resulting in the heavy version being 8 Da heavier than the light version. The corresponding isotope labeling for arginine is $^{13}C_6$ $^{15}N_4$, resulting in a mass increase of 10 Da.[1]

For triple SILAC (comparing three samples) an intermediate label is required, and a popular intermediate labeling is $^2H_4$, with a mass increase of 4 Da for lysine, and $^{13}C_6$, with a mass increase of 6 Da for arginine.

There is very little chemical difference between the different isotopes included in the amino acids, so the cells behave in (more or less) exactly the same way, irrespective of their labeling. When labeled amino acids are supplied to cells in culture, they are incorporated into all newly synthesized proteins during protein turnover, cell growth, and cell division. Over time, this incorporation becomes 100 % and hence reproducible.

Figure 12.1 illustrates a triple SILAC experiment using the labels mentioned above. Note that in an experiment one can choose to only label lysine, only arginine, or both. Labeling only one of the amino acids results in less complex spectra, but produce fewer labeled peptides, thus reducing the achievable sequence coverage.

Use of internal standards

Given that SILAC incorporates the labels metabolically it was first limited to cells in culture. However, it has since been used to label entire organisms, from fruit flies to animals as complex as mice. Analyzing human tissue is however not possible using the standard SILAC procedure, but is obtainable by using spike-in standards.

In such cases, double SILAC is used, where each experimental run contains an (unlabeled) experimental sample and a SILAC-labeled sample as a reference. The ratios between two experimental samples must then be calculated via the reference relative ratios, as shown in Chapter 9.6.

Choosing an appropriate reference sample is essential, and Geiger et al. (2010) has demonstrated that using a mix of labeled protein lysates from several (five) previously established labeled cell lines gives good results. The five cell lines together seem to be a good representative of the tissue under consideration. This use of mixed labeled cell lines is referred to as *super-SILAC*.

---

[1] Note that while we for simplicity provide the masses of the labels as integers, this will not be sufficient when used in real calculations, and more accurate masses have to be used.

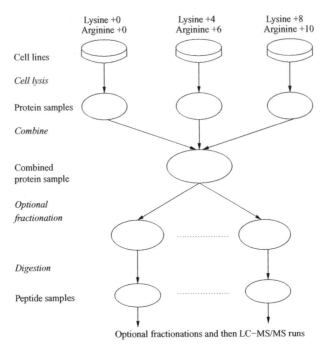

Figure 12.1   Illustration of a triple SILAC experiment.

12.1.2   Chemical labeling

From a data analysis perspective the use of the different chemical labels is all very similar. However their chemical structures are different, and there may also be differences in how they attach to the proteins/peptides.

ICAT

ICAT (Isotope Coded Affinity Tags) can be used to compare two situations or samples. The labels consist of four groups, as shown in Figure 12.2, and are incorporated at the protein level.

- *Isotope coded tag*: The group that distinguishes between the two variants of the peptide.

- *Affinity tag (biotin)*: Used to recognize and isolate the labeled peptides, thus reducing the sample complexity before LC-MS/MS analysis.

- *Cleavable linker*: Used for cleaving off the affinity tag after isolation. In this way the size of the label attached to the peptide is reduced.

- *Reactive group*: Used for covalent binding to the thiol group of cysteines in proteins.

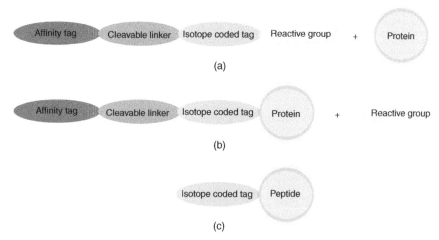

Figure 12.2   Illustration of ICAT labeling. (a) shows the label before binding, (b) is after binding, and (c) is after labeled peptide isolation, at which point the affinity tag has been cleaved off.

The composition of the tag after cleavage is $C_{10}H_{17}N_3O_3$ for the light variant, and $^{13}C_9CH_{17}N_3O_3$ for the heavy variant, with monoisotopic masses of 227.13 Da and 236.16 Da respectively.

ICAT labels are attached to cysteines only, and given that cysteine has an occurrence rate of approximately 1.5 % in the human proteome, only about 15 % of tryptic peptides will contain a cysteine and will therefore be labeled.[2] This reduction in complexity simplifies the LC-MS/MS process, but reduces the number of peptides that can be used for quantifying each protein. Note that it is also possible that certain proteins do not contain any identifiable tryptic peptides with a cysteine, thus making it impossible to identify or quantify these proteins.

Figure 12.3 shows the ICAT procedure.

ICPL

ICPL (Isotope Coded Protein Label) can be used to compare two, three, or four samples. The labels are incorporated into the proteins at free amine groups. Such free amines exist only at the N-terminus of the peptides and on the side chain of lysine.

The mass increases for the four labels are:

- $ICPL_0$: 0.00 Da;
- $ICPL_4$: 4.05 Da, H_4 is replaced by $^2H_4$;

---

[2] Note that this assumes an average peptide length of 10 amino acids, which corresponds to the amino acid frequencies of lysine and arginine in the human proteome, both of which are about 5 %, yielding a 10 % combined prevalence for tryptic cleavage sites.

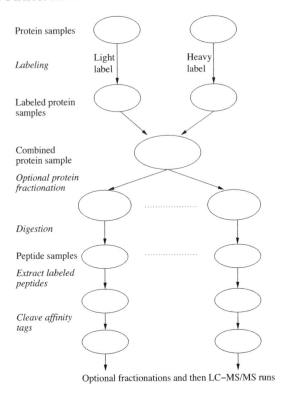

Figure 12.3   Illustration of the ICAT experimental workflow.

- ICPL$_6$: 6.02 Da, C_6 is replaced by $^{13}C_6$;
- ICPL$_{10}$: 10.07 Da, H_4C_6 is replaced by $^2H_4\,^{13}C_6$.

Dimethyl labeling

Dimethyl labeling has been used to compare two, three, or four samples. The labels are incorporated into the peptides, at the free amine groups, and the labeling is carried out by allowing formaldehyde to react with these amines. This is shown in Figure 12.4 for the light label, resulting in a mass increase of 28 Da.

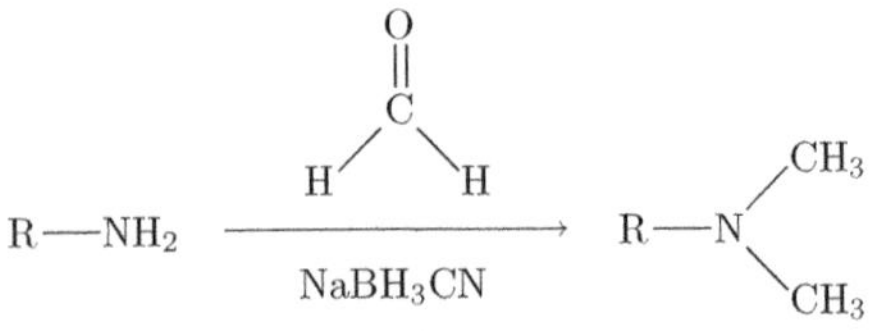

$$
\text{R}-\text{NH}_2 \xrightarrow[\text{NaBH}_3\text{CN}]{\quad \overset{\displaystyle \overset{O}{\|}}{\underset{H\ \ \ \ H}{C}} \quad} \text{R}-\text{N}\begin{smallmatrix}\text{CH}_3\\[4pt]\text{CH}_3\end{smallmatrix}
$$

Figure 12.4   The reaction used in dimethyl labeling, here shown for the light label.

Different labels can be constructed using ^2H and/or ^{13}C in the formaldehyde and/or in the reducing agent ($NaBH_3CN$).

Three-sample multiplexing can be achieved by:

- NH_2 changes to NC_2H_6, a mass increase of 28 Da (light label);

- NH_2 changes to $NC_2H_2\,^2H_4$, a mass increase of 32 Da (intermediate label);

- NH_2 changes to $N\,^{13}C_2\,^2H_6$, a mass increase of 36 Da (heavy label).

A challenge with dimethyl labeling is the reductive deamidation of peptides. This increases the hydrophobicity, causing the labeled peptides to elute later than the nonlabeled peptides.[3] Note that the problem is reduced when only the two labels with deuterium (heavy hydrogen; ^2H) are used as reagents, that is, the intermediate and heavy labels. Also note that the shift in retention time is dependent on the properties of the peptide itself, that is, the composition of amino acids, peptide length, and the overall hydrophobicity of the peptide.

12.1.3   Enzymatic labeling – ^{18}O

^{18}O labeling allows two samples to be compared. ^{18}O atoms are incorporated into the carboxyl termini of tryptic peptides in one of the two samples, whereas the other remains unlabeled. This results in a 4 Da difference between the peptide variants. Incorporation takes place during the digestion of the proteins, which is carried out in heavy water ($H_2\,^{18}O$).

The process relies on the ability of trypsin to exchange a light oxygen isotope (^{16}O) at the carboxyl terminus of a peptide, with a heavy oxygen (^{18}O) from the water in which the digestion reaction takes place. It is similar to the cleavage reaction of trypsin, only that no peptide bonds are cleaved, since the peptide already has a free carboxyl terminus. It is however important to completely shut down the activity of trypsin after the digestion step, since all subsequent steps are carried out in normal ($H_2\,^{16}O$) water and any remaining tryptic activity could thus lead to so-called back-exchange, where a heavy oxygen isotope at the carboxyl terminus is exchanged for a light oxygen isotope from the solvent.

This back-exchange issue is particularly important given that a peptide carboxyl terminus is made up of two oxygen atoms, both of which should ideally be replaced with the heavy oxygen isotope. This yields a fully labeled carboxyl function ($-C\,^{18}O_2$), and corresponds to a mass increase of 4 Da. In the case of incomplete incorporation, or partial back-exchange, only one oxygen atom may be a heavy isotope, yielding a partially labeled carboxyl function ($-C\,^{16}O\,^{18}O$) with a mass increase of 2 Da. Partial labeling is thus undesired as it creates a third peak of intermediate mass that complicates the analysis. This complication is made worse as the limited mass increase of 2 Da causes the isotope envelope of the light peptide variant to overlap with the heavy variant, confounding the calculation of separate intensities for each variant.

---

[3] Dimethylation can therefore not be used for absolute quantification.

Special care thus needs to be taken to accurately follow a reliable protocol when performing ^{18}O labeling. A good protocol can be found in Staes et al. (2004), although it should be noted that the authors later added 4M guandinium hydrochloride (GuHCl) to the step where the labels are combined. This latter addition is required to silence all remaining trypsin activity, and thus limits the last possibilities for back-exchange.

12.2   Experimental setup

The peaks that can be used for quantification depend largely on the peaks that are selected for MS/MS processing. This link with identification needs to be taken into account when determining how the experiment should be performed. As a result, there are several options for the LC-MS/MS experimental setup:

- Perform an ordinary LC-MS/MS run, and save all spectra. Standard ways of selecting peaks for MS/MS from an initial MS scan can be used, for example, analyzing the k most intense peaks. The MS spectra are then analyzed for quantification purposes, and the MS/MS spectra are used for peptide identification.

    A problem with this approach is that there is no guarantee that MS/MS spectra for all presumptive peptide pairs exist.

- First perform an LC-MS run, and save the MS spectra. Find presumptive pairs of peaks that constitute possible variants of a peptide. Note the retention time and m/z values. Then perform a new LC-MS/MS run trying to replicate the same conditions, and use the pairs determined from the previous LC-MS run to create an *inclusion list* of peaks for MS/MS. Such an inclusion list contains the m/z values that should be selected for MS/MS analysis in any given retention time window. When configured with an inclusion list, the mass spectrometer will only focus on the ions on this list, and will disregard other signals, regardless of their intensity. This second run then produces the MS/MS spectra that correspond with the peak pairs identified from the first run.

    An advantage of this approach when compared to the single-run approach is that the bias towards the most abundant peaks is partially removed. This approach is sometimes called *directed quantification*.[4]

    A problem with this approach is that it can be difficult to achieve exactly the same experimental conditions, something that can result in different retention times of the relevant peptides, different m/z values, or even the absence of previously observed peaks in the second run.

- Analyze the MS spectra in-line during the LC-MS/MS run, and use the presumptive pairs to directly instruct the instrument to select the corresponding peak (or peaks) for MS/MS analysis.

---

[4] In Domon and Aebersold (2010) the quantification methods are divided into three strategies: Discovery (or shotgun), directed, and targeted.

A problem with this approach is that the in-line analysis can take up so much time that the peaks that should be targeted for MS/MS may already have fully eluted from the chromatographic column, and are therefore missed. The speed of in-line analysis is therefore of paramount importance to the success of this approach.

12.3   MaxQuant as a model

MaxQuant is a highly popular, free tool for label based quantification using MS spectra.[5] We will therefore describe label based quantification by MS spectra based on how this procedure is handled by MaxQuant. Other tools for this purpose can be found in the Bibliographic notes. Note that MaxQuant is specifically aimed at high resolution MS data, such as Orbitrap data, and that its data input is focused on data from Thermo Fisher instruments.

It must be underlined that the presentation here is neither a full nor a perfectly accurate description of the MaxQuant software, which can be found in Cox and Mann (2008, 2009). We will however use examples and ideas from MaxQuant, but with some of the finer details left out.

An important assumption for the analysis is that the *peptide masses (MS level) can be determined with high accuracy and precision* (at least 7 ppm). On the other hand the accuracy of the fragment ions (MS/MS) can be relaxed (typically allowing an error of up to 0.5 Da).

MaxQuant was originally developed for the analysis of two samples, but now also supports the analysis of three samples. This is achieved by performing the pair-wise analysis for each of the three sample pairs. The experimental setup is very similar to the first alternative in the preceding section, analyzing the result file from a single LC-MS/MS run.

12.3.1   HL-pairs

The basic quantification task is to determine the relative abundance of two peptide variants, and then infer the relative protein abundances. Going from peak intensities over peptide abundances to (relative) protein abundances can be done in several ways, as outlined in Chapter 5.8. Figure 12.5 (a repeat of Figure 5.4) shows that a peptide variant can have numerous occurrences, rendering the calculation of the relative abundance for a peptide more complicated.

For simplicity we call the two peptide variants *heavy* and *light*. We then define an *HL-pair* (heavy–light pair)[6] as a pair of corresponding subspaces of the peptide

---

[5] Note that MaxQuant can also be used for label free quantification experiments.

[6] Note that when more than two labels are used an HL-pair can also be used to represent other pairs than heavy and light.

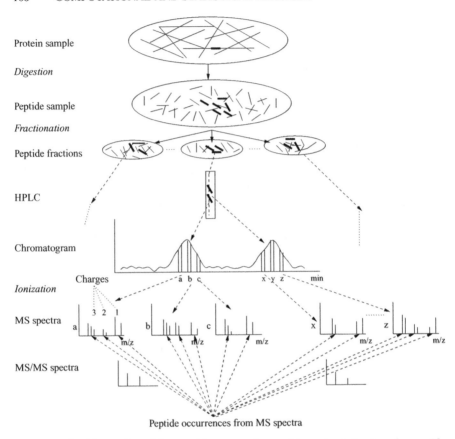

Figure 12.5   Illustration of the peptide space of a peptide variant. Copies of a specific peptide variant are highlighted.

spaces, characterized by the fact that a relative protein abundance is calculated from the relative abundances of the HL-pairs of the protein, for example, by taking the median across all these pairs.

From Figure 12.5 we see that the subspace of an HL-pair is bounded by a single set of variant peaks on the one hand, and the complete peptide space on the other hand. An example of a space between these two extremes is to collect in an HL-pair all the variant peaks of a peptide with the same charge. This is the approach used in MaxQuant, where such HL-pairs are termed *SILAC-pairs*. We will however use the more generic term HL-pair as defined above.

Note that for an HL-pair to be used in the protein quantification it must be identified (directly or indirectly) by at least one MS/MS spectrum in either one of the (heavy or light) peptide spaces.

12.3.2   Reliability of HL-pairs

The distance between the light and heavy version of an HL-pair is fixed in the m/z-scale. This distance depends on the mass of the label(s), the charge, and the number of labels attached.

In order to be certain that a valid peptide HL-pair has been found, each peak in the pair ought to be identified by an MS/MS spectrum, and the distance requirement must satisfied. This stringent requirement is an extreme case, where HL-pairs are constructed only from peaks that are positively identified as coming from variants of the same peptide.

The other extreme is to try to build the HL-pair from peaks assumed to come from the same peptide, using only the distance requirement. Assumed HL-pairs then have to be identified by at least one MS/MS spectrum from one of the peaks. MaxQuant use this latter, more permissive option.

12.3.3   Reliable protein results

Although it may be possible to analyze differential protein abundance using only the results from a single LC-MS/MS run, the developers of MaxQuant recommend that at least 10 runs of the combined sample are performed, and that the results are analyzed together in order to get reliable statistics. The analysis procedure is however the same in both cases. This means that when calculating the relative abundance of a protein the relative abundances from the HL-pairs of all the runs are used together.

12.4   The MaxQuant procedure

We will now briefly describe the MaxQuant procedure. Performing the quantification experiment can be described in four steps:

1. Recognize (presumptive) HL-pairs.

2. Estimate HL-ratios.

3. Identify HL-pairs by database search.

4. Infer protein data.

These steps are elaborated below.

12.4.1   Recognize HL-pairs

To recognize HL-pairs we should keep in mind that a peptide:

- can occur in several successive spectra;

- usually occurs as an isotope pattern;

- can occur with different charges, with each charge state yielding a separate isotope pattern.

Note that MaxQuant expends a lot of effort in obtaining peptide masses that are as accurate as possible, as will be shown in the following description.

A single peak in a spectrum is termed a 2D peak, and a set of 2D peaks from a peptide occurring in successive spectra is called a 3D peak, as illustrated in Figure 12.6. Note that a given 2D peak may be absent from the 3D peak, for example, caused by too small intensity or errors in the mass determination.

The actual step at which peaks are combined into isotope patterns depends on whether the patterns are determined using 2D peaks or 3D peaks. This may be a matter of preference, but 3D peaks do yield more information that can be used to validate the correctness of proposed pattern. The alternative is to first identify 2D isotope patterns and then combine these into 3D isotope patterns. Note that when we later in this chapter refer to isotope patterns, we are referring to the 3D patterns.

Isotope patterns and different charges, along with the concepts of 2D and 3D peaks are illustrated in Figure 12.6. The figure also shows that several isotope patterns can exist for a peptide, forming the basis for different HL-pairs (because of different charges).

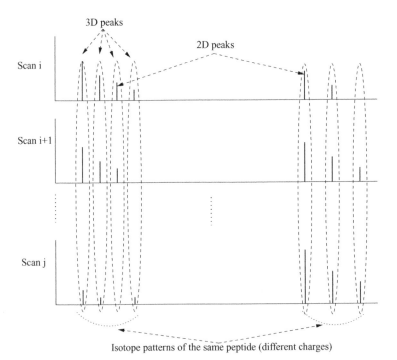

Figure 12.6   Illustration of 2D and 3D peaks, isotope patterns, and different charges. MS spectra from consecutive scans are illustrated.

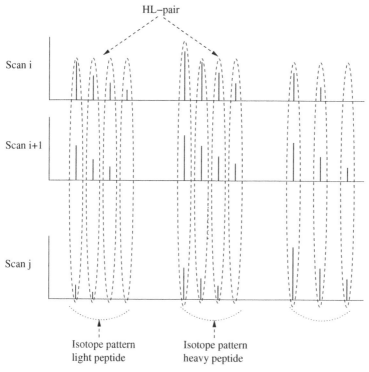

Figure 12.7   Illustration of HL-pairs.

In the procedure we locate heavy and light variants by pairing isotope patterns, as illustrated in Figure 12.7, and these paired 3D isotope patterns then constitute HL-pairs.

The relative abundances are calculated for the pairs, called *HL-ratios*. The procedure for determining HL-ratios can be described in five steps:

1. detect 2D peaks;

2. combine 2D peaks into 3D peaks;

3. determine isotope patterns;

4. determine HL-pairs;

5. recalculate the peptide m/z.

2D peak detection

2D peaks can be found by traditional methods. Due to the high precision and accuracy mandated by MaxQuant, a simple procedure can be performed: Searching for local intensity maxima. The lower and upper limits in the m/z-direction are determined, as shown in Figure 12.8 a,b. A center m/z-value is determined, and the intensity is calculated by summing the intensities of all the raw data points.

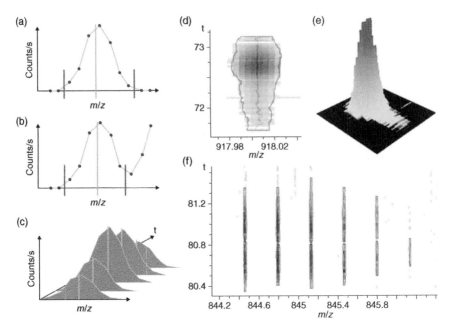

Figure 12.8 Illustration of peak detection and overlapping isotope patterns. Reprinted from Cox and Mann (2008) by permission of Nature Publishing Group.

3D peak detection

The procedure for 3D peak detection in MaxQuant is described next.

- A 2D peak in a spectrum (scan) j is connected to a peak in scan $j + 1$ or $j + 2$ if their (centroid) masses (m/z) differ by less than k ppm ($k = 7$).

- A 3D peak is defined as the maximum chain of 2D peaks connected in this way.

- There must be at least two 2D peaks in any 3D peak.

- The intensity profile of the 3D peak (composed of the different intensities of the 2D peaks, *3D peak intensity profile*) is smoothed and checked for local minima. If a minimum is found that's value is below $\frac{1}{h}$ (with h a constant) times the lower of the two local maxima in the intensity profile, the 3D peak is split in two at the minimum position.

- The center mass of the 3D peak is calculated as:

$$\bar{m} = \frac{\sum_{j=1}^{n} m_j I_j}{\sum_{j=1}^{n} I_j}, \tag{12.1}$$

where n is the number of spectra, and m_j, I_j represent the mass and intensity of the 2D peak in spectrum j.

- The standard error of the mass ($\Delta\bar{m}$) is estimated through bootstrapping over $B(= 150)$ bootstrap replicates.[7] A bootstrap replicate consists of randomly and uniformly drawing with replacement n values (b) from the indices $1, \ldots, n$, and for each bootstrap b the mass $\bar{m}_b$ is calculated using Equation 12.1. $\Delta\bar{m}$ is then calculated as

$$\Delta\bar{m} = \sqrt{\frac{\sum_{b=1}^{B}(\bar{m}_b - \bar{m})^2}{B - 1}}.$$

The standard errors are later used in the calculation of the final peptide mass.

Example Let there be $n = 6$ spectra in a 3D peak, and let the masses be 228.464, 228.103, 229.083, 228.640, 228.764, 228.526, and assume for the sake of simplicity that all the intensities are equal. Then $\bar{m} = 228.597$. Assume that a bootstrapping replicate (b) draws the indices 3, 4, 6, 3, 1, 1 (note the repeated drawing of indices 1 and 3), then $\bar{m}_b = 228.710$.
△

3D peak detection is illustrated in Figure 12.8 c–e.

Determining isotope patterns

Isotope patterns are determined from the 3D peaks, but the procedure is not straight-forward, mainly due to possible overlap between isotope patterns.

We can formulate three constraints for the 3D peaks in an isotope pattern, see Figure 12.9. In MaxQuant the high accuracy of an Orbitrap is utilized in setting the constraints.

1. The mass difference between consecutive peaks must be similar. Differences are determined using the two masses, the standard errors and the masses of ^{12}C, ^{13}C, ^{32}S, ^{34}S.

2. The mass difference has to correspond to allowed charge values.

3. The 3D peak intensity profiles should be similar for all the 3D peaks in the pattern. Pearson's correlation coefficient is used to verify this similarity, see Section 20.8.1.

4. The isotope pattern intensity profile should correspond to the theoretical intensity profile of a peptide with mass corresponding to the actual m/z and charge.

Note that the intensity profile of a pattern is shown along the m/z-axis, while the 3D peak intensity profile is shown along the retention time axis, as illustrated in Figure 12.9.

---

[7] Bootstrapping is a technique used to iteratively improve a classifier's performance.

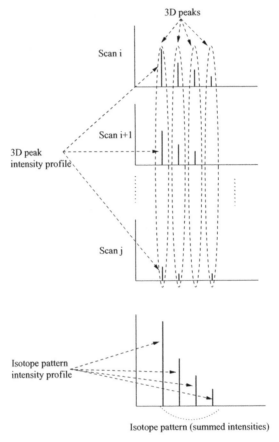

Figure 12.9   Illustration of a 3D peak intensity profile and an isotope pattern intensity profile.

Search for isotope patterns   An appropriate procedure for revealing the patterns is to use graphs, a strategy also used by MaxQuant:

- the nodes are the 3D peaks;

- there is an edge between two nodes if they can be neighbors in a pattern (satisfying the pair-wise constraints above).

Connected subgraphs then represent potential isotope patterns. It is however important to verify that all pairs are consistent (specifically, that all of the peaks carry the same charge). A connected subgraph is then decomposed into consistent subgraphs. This can be done by first detecting the largest consistent subgraph, and then recursively finding the largest in the remaining data.

This procedure ensures that a consistent pattern cannot have gaps, meaning that a pattern must consists of successive isotopes.

Determine HL-pairs

In principle each pair of isotope patterns should now be tested to see if they could constitute an HL-pair. The constraints for two isotope patterns to form an HL-pair are:

1. Their charge (z) should be equal.

2. The distance between the corresponding 3D peaks in the two patterns should be in accordance with the charge and the known label masses. If the mass difference between the heavy and the light labels is a constant d, then the allowed distances are

$$\frac{nd}{z}; \; n \leq n_m,$$

   where n is the number of labels attached to the peptide, and n_m is the maximum number of labels that are considered (an equal number of labels is assumed for each version).

   If the mass difference can vary, as is the case for SILAC (where the labels attached to lysine and arginine are different), the number of allowable differences can increase considerably. However, if the number of labels per peptide is restricted to one (representing the ideal case for SILAC when used on tryptic peptides), this increase remains manageable.

   It may also be useful to verify that the number of attached labels n, is actually chemically possible for the peptide sequence under consideration. This is however only possible if the peptide is identified, since the number of possible labels is sequence dependent.

3. The 3D peak intensity profiles of the two patterns should be similar. This can be verified in the same way as in the test for finding isotope patterns, here using the sum of the isotopes over the isotope pattern, as illustrated in Figure 12.10 where the HL-pairs are contained in three MS spectra. The two summed 3D peak intensity profiles are then compared.

   Note that MaxQuant also tests whether the correlation increases if one of the intensity profiles is shifted by one scan, due to the fact that the labeling can have a minor impact on the retention time.

Calculating more accurate peptide masses

All the recognized m/z-values for the 2D peaks of the (assumed) same peptide can now be used in conjunction with the peptide charge (determined from the isotope pattern) to calculate an (accurate) *mass* of the peptides. First the mass of each 2D peak is determined, and then a weighted average is calculated, where the intensities of the peaks are used as weights. The standard error of $m_{peptide}$ is again calculated by bootstrap resampling of the 2D peaks that go into the calculation. Remember that there are multiple peaks for each HL-pair.

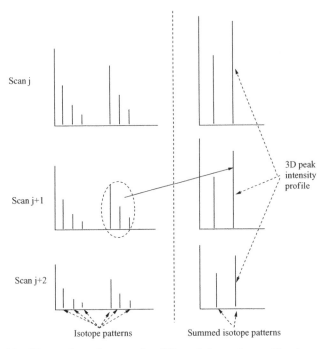

Figure 12.10   Illustration of comparing 3D peak intensity profiles in an ideal case.

Nonlinear mass recalibration   If HL-pairs occurring with more than one charge state are recognized, nonlinear mass recalibration can be performed. In MaxQuant two HL-pairs are detected as a *charge pair* if:

- the retention time profiles correlate (and they should indeed be very similar, since ionization happens after chromatographic separation);

- the two peptide mass estimates are the same within a given threshold (7 ppm).

Recalibration for more exact mass estimation is then performed using the masses of all the HL-pairs of the (assumed) same peptide. The estimates of the peptide mass errors are also adjusted. How this is done is detailed in Cox and Mann (2008, 2009).

12.4.2   Estimate HL-ratios

The abundance ratio between the heavy and light versions of an HL-pair can now be estimated from the ratios of all the corresponding 2D peaks that are present in both versions. In Figure 12.10 there are nine such pairs. Again the formulas in Chapter 5.8 can be used, and in MaxQuant the method of fitting a linear regression line is used. Note that ratio estimation is done *before* any identification, at which stage there is no verification that the pairs are actually derived from peptide variants.

Experimental normalization of the HL-ratios

The standard assumption that only a few of the proteins are differentially abundant is employed when normalizing the HL-ratios. Usually the median of the logarithm of the ratios is set to zero, and the other ratios are corrected accordingly. For MaxQuant specifically, we can add the following:

- The normalization is done separately for different intensity bins, maximizing the effect of decreasing variation at increasing intensity.

- Normalization is also performed separately for lysine and arginine labeled peptides to compensate for any label-specific bias.

12.4.3   Identify HL-pairs by database search

The obtained HL-pairs can now be identified, that is, it can be determined which peptide (derived from which protein(s)) is the origin for each of the HL-pairs. Also any isotope patterns that are not part of any HL-pairs (unknown state) should be identified, such that possible partners can be searched for in the spectra, given that the positions where they should occur is now known. In this way the number of HL-pairs can be increased by performing a less stringent second pass analysis.

The procedure for identification and processing of the results can be described in four steps:

1. Prepare the spectrum files for database searching.

2. Perform the search.

3. Filter identified peptides.

4. Discover new HL-pairs.

Prepare for database searching

In MaxQuant the MS/MS spectra are prepared for database searching as follows:

- Low-resolution MS/MS spectra recorded in centroid mode are filtered such that only the k most intense peaks per 100 (m/z) units are retained, $k = 6$ is the default.

- High-resolution MS/MS spectra are deisotoped and transformed to single charged peaks before the ten most intense peaks are retained.

- Three msm-files are created for each LC-MS/MS run, all with both MS as well as MS/MS spectra, one for each of the states: Heavy, light, and unknown (not yet part of an HL-pair). This is illustrated in Figure 12.11.

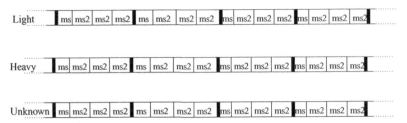

Figure 12.11   Example of the generated msm-files for one LC-MS/MS run, here with three MS/MS scans for each MS scan.

Database searching

Since the state of the precursor[8] of each MS/MS spectrum is known it can be used for searching:

- If the precursor is part of a heavy isotope pattern the labels can be set as fixed modifications (when using SILAC, these modifications are on lysines and arginines).

- If the precursor is part of a light isotope pattern and the light versions have no labels, no special actions are taken for the search. If there are light labels, then these are set as fixed modifications.

- If the state of the precursor is unknown, all labels are set as variable modifications.

The search results in several peptides that are potential matches to a spectrum. These peptides are called *peptide candidates*. This is illustrated in Figure 12.12, showing that there can be numerous MS/MS spectra for an isotope pattern, and several candidates for each MS/MS spectrum (note that the number of peptide candidates can also be zero).

Mascot has previously been used as the search engine for MaxQuant, but recently a new search engine, Andromeda, Cox et al. (2011), was developed and integrated into MaxQuant replacing Mascot.

Filter peptide candidates

After database searching the candidates have to be evaluated and filtered based on different criteria:

1. Filter peptide candidates inconsistent with the label state.

2. Filter peptide candidates based on calibrated mass data.

3. Filter peptide candidates based on identification significance.

---

[8] Here, the precursor is the 2D peak in the MS spectrum that is selected for MS/MS processing.

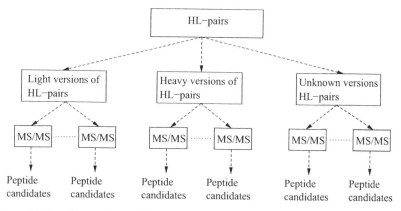

Figure 12.12   Illustration of the peptide candidates found by database searching.

Filter peptide candidates by label state information   The sequence of a peptide candidate is known, and the consistency between the sequence and the label state can be controlled. In MaxQuant the masses and charges are used to determined the number of labels if the HL-state is known, and this number is compared to the number of labeling sites in the sequence. Peptide candidates for MS/MS spectra of unknown state are filtered if not all or none of the lysines and/or arginines in the sequence are labeled.

Filter peptide candidates by mass after recalibration   MaxQuant uses candidates with an identification score above a specified threshold to calculate a possible global mass correction:

$$\Delta ppm = \frac{\sum_j \frac{m_j^2}{\Delta m_j^2} \Delta ppm_j}{\sum_j \frac{m_j^2}{\Delta m_j^2}},$$

where

- Δppm_j is the difference between the experimentally measured peptide mass m_j and the theoretically calculated mass for that peptide;

- Δm_j is the previously estimated standard error.

Δppm is then subtracted from all measured peptide masses, and all peptide candidates with a difference larger than four times the standard error between the corrected mass value and the theoretically calculated mass value are discarded.

Filter peptide candidates based on identification significance   Identification significance is calculated by first constructing a decoy database, see Chapter 4.7. The decoy database is constructed by reversing the sequences in the original database. However, a problem with this approach is that it creates peptides that have exactly the same mass composition as the peptides in the forward database (this happens in half of the cases with tryptic peptides). Thus the reverse database overestimates the number of random hits. This is a problem for MaxQuant given its reliance on the high precursor accuracy. Each lysine and arginine is therefore swapped with its preceding amino acid in the reversed sequences:

Sequence:	*MVSTRLCWKLMN* → *MVSTR* | *LCWK* | *LMN*
Decoy:	*NMLKWCLRTSVM* → *NMLK* | *WCLR* | *TSVM*
Swapped:	*NMKLWCRLTSVM* → *NMK* | *LWCR* | *LTSVM*

The decoy database thus has the same mass and amino acid distribution as the original database, but avoids the *exact* same mass values.

In MaxQuant a Posterior Error Probability (PEP) can be calculated for each identification that estimates the probability for that identification to be a false identification. For this it relies on the results from the decoy database and Bayes theorem as explained in Cox and Mann (2008).

The False Discovery Rate (FDR) is then determined, based on the identifications sorted on their calculated PEP values. Filtering can be performed on the PEP or FDR level:

- PEP: Discard all peptides with a PEP above a defined threshold;

- FDR: Accept all peptides until a predefined FDR-threshold is reached (typically 0.01).

It should however be mentioned that filtering through FDR and PEP requires the use of data from several (all) LC-MS/MS runs.

Finding secondary peptides

A single MS/MS spectrum can be derived not only from the peptide corresponding to the peak selected for MS/MS processing, but also from other peptide(s) with similar m/z values that happen to fall in the same selection window. Such so-called chimeric spectra can be dealt with in at least two ways:

- extend the identification process to also consider the case that a single MS/MS spectrum can represent more than one peptide;

- when a peptide is identified, subtract the peaks corresponding to this peptide from the spectrum, and perform a new search with a derived spectrum that retains only the remaining peaks.

Discover new HL-pairs

Many of the isotope patterns with unknown state, that is, not identified as part of an HL-pair, will have MS/MS spectra with identified peptide candidates. The state of these spectra can be determined based on the peptide sequence and the mass. One can then calculate at which mass(es) the potentially missing HL-partner is expected and what the intensity profile ought to look like, and then check if there is a corresponding isotope pattern at that location.

12.4.4   Infer protein data

To calculate the protein ratios we now have the HL-pairs from all the runs that are identified by one or more significant peptide candidates. These peptide candidates are the basis for calculating the protein ratios, which can be performed in five steps:

1. Choose the peptide candidate for an HL-pair.

2. Infer the proteins for each HL-pair.

3. Calculate a protein false discovery rate.

4. Calculate protein ratios.

5. Calculate protein ratio significance.

Choose the peptide candidate for an HL-pair

From Figure 12.12 we see that an HL-pair can be identified by several MS/MS spectra, and these spectra can have different top scoring (meaning identified by the highest significance) peptides. In MaxQuant, after the filtering, this does not often occur, and when it does the peptide with highest significance is chosen as the correct peptide candidate. This means that there is always only one selected peptide candidate for an HL-pair.

Infer the proteins for the HL-pairs – the protein inference problem

A peptide candidate can occur in several proteins, and it is often not obvious which of these is (or are!) the origin of the peptide in the sample. The assigning of peptides to proteins is known as the protein inference problem, and we here briefly explain how it is dealt with in MaxQuant.

The proteins containing the peptide candidates are organized into *protein groups*, defined as follows:

- there is one dominant protein P;

- all peptide candidates of all the other proteins in the group are also candidates for P;

- P is therefore able to explain all the peptide candidates contained in the protein group by itself;

- all possible groups are formed;

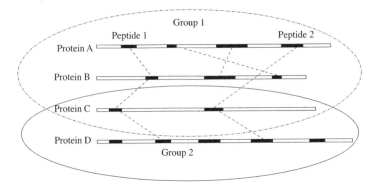

Figure 12.13   Example of two protein groups. Protein A is the dominant protein in group 1, and D in group 2. Protein C is a member of both groups, since the two peptides 1 and 2 are included in both A and D.

- a peptide candidate is called *unique* to a protein group if it occurs only in this group (and nowhere else in the considered proteome);

- a nonunique peptide candidate is (tentatively) assigned to the protein group with the highest number of peptide candidates (using Occam's razor).

An example is shown in Figure 12.13.

Calculate protein FDR

A false discovery rate is calculated for each protein group as follows:

- A PEP is calculated for each protein group by multiplying the PEPs from each of the peptide candidates. The PEP for a peptide candidate is taken from the MS/MS spectrum with the lowest PEP (remember that a specific peptide candidate can be identified from several MS/MS spectra).

- The PEP is only used for sorting the protein groups (it is not considered to be a reliable statistic beyond this ranking ability).

- FDR is determined using a decoy database created by the standard approach.

- All protein groups without unique peptides are removed.

- Proteins with only one unique peptide may be discarded (or manually examined).

Calculate protein ratios

A protein ratio is calculated as the *median* of *all* HL-pair ratios that belong to peptide candidates inferred to the protein. The median is used because it is more robust against outliers.

Normalized protein ratios are calculated using the median of the normalized HL-ratios, see Section 12.4.2.

Calculate ratio significance

As explained in Chapter 6.6.3 the distribution of the log of the ratios of the nondiffer-entially abundant proteins are expected to be a normal distribution, and differentially abundant proteins can be detected as outliers in this distribution. MaxQuant proposes a procedure where the distribution is made up of all the normalized protein ratios. The 15.87, 50, and 84.13 percentiles can then be used to achieve a robust and asymmetrical estimate of the number of standard deviations that a ratio is away from the mean, as described in Chapter 8.2.1. The significance calculated in this way is called significance A.

It is well known that the width of the bulk distribution of logarithmic ratios depends on the protein intensity. For highly abundant proteins the statistical spread of unregulated proteins is much more focused than for low abundant ones. To handle this heterogeneity of variance another significance B is proposed. It is calculated in the same way as A, but on protein subsets by grouping them into intensity bins. The proteins are divided into bins of equal occupancy such that each bin contains at least 300 proteins.

Significance A and B are then both corrected for multiple hypothesis testing, see Chapter 8.5.

12.5 Exercises

1. Show that approximately 15 % of the tryptic peptides will be labeled by ICAT labeling.

2. From a given SILAC run (with only lysines labeled) we have four successive spectra termed A, B, C, and D:

```
A:
( 508.44,   602) ( 561.18,   418) ( 561.72,   420) ( 562.23,   210)
( 563.67,   907) ( 566.23,   835) ( 566.70,   814) ( 567.27,   401)
( 802.44,   809) ( 993.34,   556) ( 994,39,   390) (1121.45,  1242)
(1122.48,  1296) (1123.40,   623) (1131.49,  2396) (1132.51,  2423)
(1133.37,  1198) (1202.28,   480)

B:
( 561.27,   628) ( 561.78,   654) ( 562.28,   302) ( 566.29,  1273)
( 566.77,  1298) ( 567.34,   612) ( 802.51,   398) ( 993.39,  1024)
( 994.44,  1056) (1121.51,   914) (1122.56,   912) (1123.47,   439)
(1128.55,   390) (1131.56,  1832) (1132.58,  1856) (1133.43,   908)

C:
( 561.23,   806) ( 561.74,   824) ( 562.26,   402) ( 566.23,  1625)
( 566.74,  1656) ( 567.25,   804) ( 802.54,   908) (1121.44,   619)
(1122.47,   654) (1123.41,   312) (1131.46,  1242) (1132.54,  1279)
(1133.39,   618) (1187.34,   412)
```

```
D:
( 561.15,   502) ( 561.71,   548) ( 562.31,   249) ( 566.18, 1012)
( 566.74, 1850) ( 567.23,   538) ( 802.52, 1212) ( 986.56,   501)
(1121.47,   490) (1122.46,   503) (1123.42,   239) (1131.52,   712)
(1131.99,   736) (1132.53,   351) (1133.01,   745) (1133.46,   491)
```

We assume that the charge is either one or two.

To solve this exercise implementing a simple program is recommended.

(a) Try to determine 3D peaks. Use a threshold of 0.1 for the m/z difference between peaks of succeeding peaks. Should some of the peaks be split into two when using $h = 2$?

(b) Calculate the mass centers of the 3D peaks.

(c) Explain how standard errors can be estimated.

(d) Construct the graph for determining isotope patterns.

(e) Determine isotope patterns.

(f) Determine HL-pairs.

(g) Comment on and estimate relative abundances.

12.6 Bibliographic notes

Software For other programs we refer to Vaudel et al. (2010), and also mention ASAPRatio (Automated Statistical Analysis on Protein Ratio), Li et al. (2005) where the peptides are identified first, and the ratios calculated second.

MaxQuant Cox and Mann (2008, 2009); Cox et al. (2009), Cox et al. (2011).

SILAC Geiger et al. (2010), Geiger et al. (2011).

ICAT Gygi et al. (1999).

ICPL Schmidt et al. (2005).

Dimethyl labeling Hsu et al. (2003), Hsu et al. (2006), Boersema et al. (2008), Aye et al. (2012).

^{18}**O** Yao et al. (2001), Reynolds et al. (2002), Liu et al. (2010).

13

Label free quantification by MS spectra

Label free quantification by MS spectra has a lot in common with label based quantification by MS spectra. The MS level ion currents are used for quantification, and the procedures used are quite similar. The main difference between the two approaches is that for label free approaches the spectra used to recognize the variants of the same peptide are derived from different LC-MS/MS runs. This means that it is not only the m/z values that must be matched between peptide variants, but also the retention times.

13.1 An ideal case – two protein samples

Suppose that we want to compare the protein abundances in two samples. We then perform separate LC-MS/MS runs for each of the samples, and record the detected spectra. Similarly to label based quantification, the basic operation here is to find corresponding peaks in the spectra, that is, peaks from the two variants of the same peptide.

In an ideal case, without noise and with perfect reproducibility, this is straightforward since the retention time and m/z values of the two variants will be equal. The corresponding MS spectra are thus found at equal retention times, and the corresponding peaks in the spectra at equal m/z-values. Abundances can then, for example, be determined using the AUC (Area Under the Curve). This idealized procedure is illustrated in Figure 13.1. If MS/MS spectra have been recorded for any of the peptide peaks these can be used for peptide identification.

Computational and Statistical Methods for Protein Quantification by Mass Spectrometry,
First Edition. Ingvar Eidhammer, Harald Barsnes, Geir Egil Eide and Lennart Martens.
© 2013 John Wiley & Sons, Ltd. Published 2013 by John Wiley & Sons, Ltd.

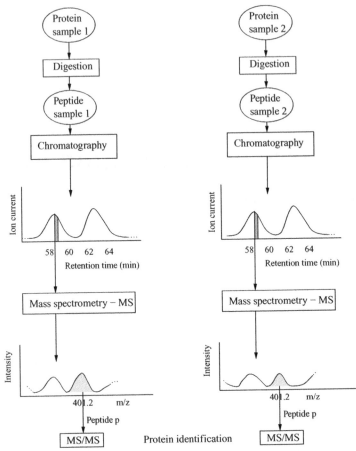

Figure 13.1   An ideal situation for the determination of relative abundance using label free MS quantification. The areas under the curve (AUCs) in the two MS spectra represent ions from peptide p in the two samples.

13.2   The real world

In practice however, the reproducibility across runs in LC-MS/MS experiments is limited, and a number of factors complicate the determination of peak correspondence across the different runs:

1. the spectra contain varying degrees of noise;

2. there can be overlapping peptide peaks in each run;

3. the retention time for a peptide varies across the runs, and this variation is nonlinear along the retention time range;

4. there can be (minor) variations in the m/z-values across the runs;

5. there can be differences in the measured intensities due to experimental factors;

6. peaks may be absent from spectra for some of the runs, due to (amongst other things) variations in ionization.

Point 1 and 2 relate to the detection of peaks, points 3 and 4 to peak matching and retention time alignment, and point 5 to (experimental) normalization. Finally, point 6 is related to the peaks that can be used in the quantification. These topics are elaborated below.

13.2.1   Multiple samples

As previously explained several replicates (biological and/or technical) are required to perform sound statistical analysis. In practice the results from several LC-MS/MS runs are therefore analyzed together, often from two situations.

13.3   Experimental setup

The three experimental setup options described in Chapter 12.2 also apply here. Mainly due to the low reproducibility of (especially) the retention time, most of the existing approaches generate the MS and MS/MS spectra in the same runs. A reasonable balance for cycling between alternating MS and MS/MS scans must therefore be found.

Since both MS and MS/MS spectra are available for the analysis, the order of the quantification and the identification has to be decided. However, doing the identification first involves numerous unnecessary identifications, since in this case we are only interested in the differentially abundant proteins, thus limiting our focus to quantifiable proteins. On the other hand, doing the identification first substantially increases the probability that peaks that are assumed to be corresponding really are derived from the same peptide.

The majority of the available software first detects corresponding peaks (or a set of peaks) that are differentially abundant, then tries to identify these, and finally infers the proteins. Note however that many of the programs restrict their function to finding peaks that are differentially abundant, and then leave it up to the user to identify the peptides and infer the proteins.

13.4   Forms

In order to obtain the quantification data we must compare the results from all the different runs in such a way that we are able to recognize corresponding *forms* across the runs. We use the term 'form' in this presentation, but other terms are also used in the literature.

A form can be, amongst other things, a single 2D peak, a 3D peak, an isotope pattern, or the results obtained after charge deconvolution. Forms are described by features, of which the most common are:

- m/z;

- retention time;

- intensity/abundance;

- charge;

- sequence (if identification is performed).

Note that the values of the features must be given in accordance with the exact definition of a form. If it is an isotope pattern, either monoisotopic or centroid mass is used, and the abundance is typically given as the sum of the intensities of all the peaks in a form.

A *form-tuple* is a set of corresponding forms from the different runs, supposed to come from different variants of the same peptide. Figure 13.2 illustrates this, with form here defined as an isotope pattern.

Note that a form-tuple can be considered as a generalization of the HL-pair concept introduced in Chapter 12.3.1, in that it can include multiple samples and/or situations.

13.5   The quantification process

We can now consider the quantification process as consisting of four tasks, which will be elaborated in the following sections.

Form detection   The forms are detected for each run separately.

Retention time correction   The retention time for the same peptide can vary across the runs, and this deviation must be corrected for.

Form-tuple detection   The corresponding forms across the runs must be determined. This is often called *aligning* the runs.

Statistical analysis

In most tools the retention time correction is included as a part of the form-tuple detection, but for optimal clarity it is best to describe it as a separate step.

Note that it is also possible to use raw data spectra for form-tuple detection, but the most common approach is to first detect forms in each run and then try to find corresponding forms across the runs. One of the reasons for the popularity of the latter approach is the large amount of data included if raw data spectra are used.

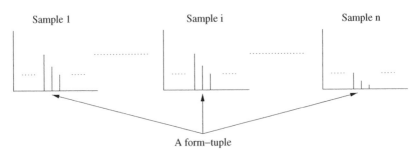

Figure 13.2   Illustration of a form-tuple; form is here defined as an isotope pattern.

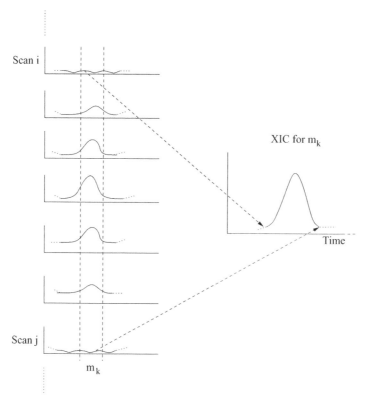

Figure 13.3   Illustration of how a 3D peak is constructed as an XIC derived from a slice across set of MS spectra.

Pair-wise alignment, consisting of retention time correction and form matching, forms the basis for multiple alignment. After first describing form detection, we will describe pair-wise retention time correction and form matching.

13.6   Form detection

As an example of form detection we describe the peak detection used in the program XCMS (Various forms of Chromatography Mass Spectrometry), Smith et al. (2006). XCMS is freely available open-source software.

XCMS uses 3D peaks as forms, with m/z, retention time and abundance as features. 3D peaks must therefore first be detected. This can be done by dividing the m/z range into slices and constructing extracted ion chromatograms (XIC[1]) for each mass slice. The 3D peaks are then peaks in the XICs. This is illustrated in Figure 13.3 for a slice m_k, where only one 3D peak is shown. Generally there can be a lot of 3D peaks in an XIC for a given mass slice.

---

[1] In XCMS an XIC is called EIBPC – Extracted Ion Base-Peak Chromatogram.

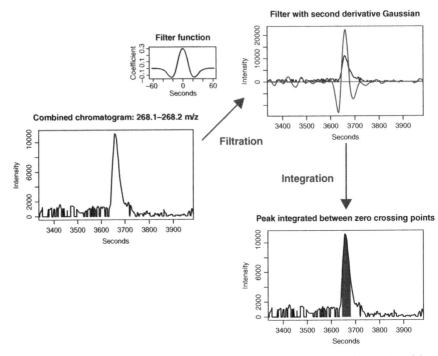

Figure 13.4   Illustration of the 3D peak detection process in XCMS. Reprinted from Smith et al. © 2006 by permission of American Chemical Society Publications.

Note that this way of detecting 3D peaks is different from the method used in MaxQuant where the 3D peaks are detected based on previously determined 2D peaks.

Now we can describe the process for determining the 3D peaks in XCMS in more detail:

1. Define m/z slices (default width is 0.1 m/z units) along the m/z range and construct an XIC for each slice. An example for m/z [268.05, 268.25] is shown on the left in Figure 13.4. The XIC is constructed for each time point by taking the maximum observed signal in the slice for that spectrum. XICs from overlapping slices are constructed in order to handle 2D peaks that are spread across more than one slice.

2. To detect chromatographic peaks a Gaussian model of the peak shape is defined.[2] This model is used as a filter to recognize peaks, as shown in the upper right of Figure 13.4. Note that the model also includes negative values.

3. Peaks are now selected using a signal-to-noise ratio cutoff.

---

[2] Actually a second-derivative Gaussian model is used, and 13 seconds is used as the standard deviation.

4. The intensity of a peak is determined by integrating the unfiltered XIC between the zero-crossing points of the filtered XIC, as shown in the bottom right of Figure 13.4. Note that any background intensity is thus included in this calculation.

13.7   Pair-wise retention time correction

After form detection we have a set of forms for each run, and the next goal is to find corresponding forms (form-tuples) across the runs by aligning these forms. We know that the retention time of a peptide can vary across the runs, and one way of handling this is to correct the retention time for each run. This is done as part of the aligning process for finding form-tuples, and how this is performed depends on the alignment approach. The underlying idea is however common: Perform the correction of each run relative to another 'run.' We will therefore briefly describe this basic pairwise correction before going into the different ways of aligning.

We put the other run in apostrophes to underline that it can be a real run, or constructed from several runs (e.g., medians), or constructed in other ways. We call the constructed 'run' the *scheme*. Later we will see that the scheme can be modified during the multiple alignment process.

Note that the retention time is a feature of the forms. One way of performing the corrections is therefore to try to recognize corresponding forms between the run and the scheme.

13.7.1   Determining potentially corresponding forms

Internal standards can be used for the detection of corresponding forms. Without internal standards corresponding forms can be determined using a rough allowed deviation of, for example, 5 minutes for the retention times, and 0.05 for the m/z values, depending on the mass accuracy of the instrument. All form-pairs found in this way are considered as potentially corresponding forms, and some of these are collected for subsequent use in a follow-up retention time correction.

The methods used to determine which form-pairs that can be used for this purpose have to consider several issues:

- Most of the proteins will have equal abundance in the two runs. We can therefore use forms related to these proteins. If the spectra are intensity normalized we can search for forms with roughly matching retention time and m/z as defined above, but also with similar abundances.

- High abundance forms are more reliable (with 'high' for example defined as higher than the median intensity).

- A form in one run may have several corresponding forms in other runs. This can be taken care of in the correction process.

Table 13.1   Data for a linear correction example.

Input	r_i	40.0	70.0	100.0	120.0	150.0
	s_i	35.0	60.0	90.0	105.0	140.0
Results	d_i	5.0	10.0	10.0	15.0	10.0
	d_i^*	7.0	8.5	10.0	11.0	12.5
	r_i^*	33.0	62.5	90.0	109.0	137.5

13.7.2   Linear corrections

Assume that we have found n potentially corresponding forms between the currently selected run and the scheme. Denote the retention time of the run forms as $\{r_i\}$, and of the scheme as $\{s_i\}$. We then calculate the correction for each r_i using the distances to s_i.[3]

First we assume that the correction is a linear function of the r_is. A procedure using linear regression, as described in Section 20.9, is used:

1. For every i calculate $d_i = r_i - s_i$.

2. Define a regression line $d_i^* = a + br_i$ such that $\sum(d_i - d_i^*)^2$ is minimized.

3. Calculate corrected retention times $r_i^* = r_i - d_i^*$.

Example Assume that we have the values $\{r_i\}$ and $\{s_i\}$ and calculate $\{d_i\}$ as shown in Table 13.1. The values are illustrated in Figure 13.5.

The regression line is approximately calculated as $d_i^* = 5 + \frac{1}{20}r_i$, as shown in the figure, and the corrected retention times are then easily calculated resulting in the values found in the final row of the table.
$\triangle$

To simplify the problem that several forms in the scheme can correspond to the same form in the run (and vice-versa) one can search for outliers and remove these before the corrections.

13.7.3   Nonlinear corrections

It is well known that the shift in retention time is not linear over the whole retention time range, and nonlinear corrections are therefore most often used, for example, LOWESS, see Chapter 7.6.

A piece-wise linear warping procedure can also be employed, Podwojski et al. (2009). This is performed by dividing the chromatographic range into several segments, with linear correction performed for each segment individually. The number of segments must be equal in the two chromatograms.

---

[3] In quantification literature the retention time correction is often called *dewarping*.

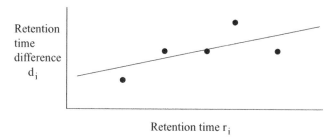

Figure 13.5   Illustration of linear retention time correction.

If two corresponding segments in the two chromatograms have a different number of forms, the smallest segment is increased by linear interpolation. The correlation between two segments can be measured by Pearson's correlation coefficient. The problem can then be formulated as an optimization problem by finding the set of segment pairs where the sum of the correlation coefficients is maximized. This optimization problem can be solved by a dynamic programming algorithm.

In Mueller et al. (2007) the problem of multiple potentially corresponding forms is solved in an iterative procedure.

13.8   Approaches for form-tuple detection

Form-tuple detection can be considered as a multiple alignment problem, which is a well studied problem in other areas of bioinformatics. The methods are often divided into four types:

- using a reference for alignment;
- complete pair-wise alignment;
- hierarchical progressive alignment;
- simultaneous iterative alignment.

The four approaches are illustrated in Figure 13.6 for a set of LC-MS/MS runs.

The first three approaches construct the multiple alignment by constructing sub-alignments (alignments of less than n runs, n being the number of runs to be aligned), and the (sub)alignments are constructed by performing a set of pair-wise alignments. Before describing the methods for multiple alignments we will therefore first briefly describe pair-wise alignments, also referred to as *pair-wise matching*.

13.9   Pair-wise alignment

Pair-wise alignment starts with two lists of forms $x = \{x_1, \ldots, x_n\}$ and $y = \{y_1, \ldots, y_m\}$. An alignment can be considered as a list of pairs of matching forms, each pair being one form from each of the lists. The task is to find the 'best' list of

Use of a reference   Complete pair-wise alignment

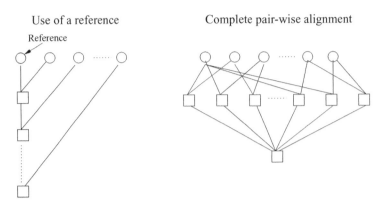

Hierarchical progressive alignment   Simultaneous iterative alignment

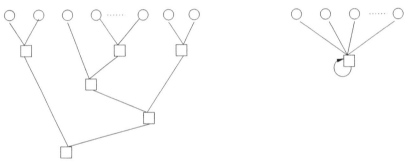

Figure 13.6   Illustration of the four different approaches for multiple alignment. Circles represent form lists, and squares represent (sub)alignments.

alignments, while simultaneously maintaining the retention time order of the items in the lists.

Maintaining the retention time order is a reasonable constraint given that the peptides are separated by the same physical properties in all runs. What is meant by 'best' can vary, but will depend on the similarity between the forms in the list. We therefore have to use methods for measuring the distance or the similarity between forms in the two lists.

The matching of two sets of forms has a lot in common with other alignment challenges in bioinformatics, for example, the alignment of protein sequences, protein structures, or gels from protein separation by 2D SDS-PAGE. Note however, that not all forms have to be included in the form alignments considered here; in other words, the alignment does not have to be complete.

13.9.1   Distance between forms

Remember that a form consists of several features, such that each form x_i (y_j) defines a vector of values $v_{i,k}(v_{j,k}), k = 1, \ldots, m$, where m is the number of features, and k can represent m/z, retention time, charge, intensity, and so on.

Proximity methods described in Section 21.1.1 can be used. Another simple similarity measurement consists of defining thresholds for the maximum allowed differences for each feature in the comparison, and let the similarity be 1 if all the differences are within these thresholds, and 0 otherwise. In this binary classification system, only pairs with a similarity of 1 will be in the alignment.

In Voss et al. (2011) a distance (which is essentially a revised Mahalanobis distance) is defined as:

$$d(x_i, y_j) = \Psi((x_i - y_j)^T W(x_i - y_j)),$$

where $\Psi(\psi) = \psi$ if $\psi \leq 1$, and $\Psi(\psi) = \infty$ for $\psi > 1$. The weight matrix W is a diagonal matrix with diagonal values $1/t_k^2$, where t_k is a user-defined threshold for the shift tolerance for feature k. These threshold values must be set in accordance with the precision of the instruments used. We see that this distance measurement takes into account the different magnitudes of the feature values.

Example We only consider the features m/z and Rt, and define the thresholds as $t_{m/z} = 0.1, t_{Rt} = 8$. Let $x_i = (600.1, 470)$, $y_j = (600.2, 476)$, then the value inside Ψ becomes 1.54, hence $d(x_i, y_j) = \infty$. If $x_i = (600.15, 460)$ $d(x_i, y_j) = 0.75$.
△

Note that the threshold for a feature is usually independent of the feature values. It might however be more reasonable to have them depend on the values. For instance, the m/z error tends to vary with the measured m/z, such that larger masses have a higher errors. m/z (and mass) errors are therefore often formulated in relative terms as ppm errors.

13.9.2   Finding an optimal alignment

Finding an optimal match can be achieved in several ways, of which we will here describe three. Remember that maintaining the retention time order is usually required. Also note that what is meant by optimal can vary, for example, some emphasize a high number of matches, while others put the most stress on the similarity of matching forms.

Using least distances

In Lange et al. (2007) the following procedure is proposed:

1. Two forms can be matched only if both are each other's best match (least distance).

2. For each of the forms, the distance to the second-best match is significantly larger then the distance to the best match.

In this way each form is either matched uniquely to a form in the other list, or not matched. Also, there is not necessarily a threshold for the distances between forms being matched.

Two step matching

In SuperHirn, Mueller et al. (2007) a two step matching procedure is used. In the first step, forms are matched if the differences between the features satisfy the tolerances. Note that this means that the matching can include ambiguities, which then have to be resolved in a second step.

> **Using dynamic programming**   When a score for each pair of forms in the two lists has been calculated, an optimal alignment can be constructed using dynamic programming in the same way as when protein sequences are aligned. The score between two forms can, for example, be defined as the inverse of the distance between them ($\frac{1}{d(x_{i,k}, y_{j,k})}$).

Gaps should yield a score of zero. This is in principle the *dynamic time warping* procedure used in Prince and Marcotte (2006).

13.10   Using a reference run for alignment

One of the LC-MS/MS runs is selected as a reference and becomes the current alignment. The other runs are then iteratively selected as the current run and aligned to the current alignment. Each iteration cycle consist of two main tasks:

1. Correct the retention time of the current run.

2. Extend the current alignment by the current run.

Thus $n - 1$ iteration cycles are performed to align n runs. This approach works well if a reference can be found for which the results are of very good quality. However, forms not detected in the reference cannot always be incorporated in the alignment, though this depends on how the two main tasks are implemented.

We here sketch an implementation partly based on Lange et al. (2007).

- The retention time of the current run l is corrected relative to the reference, resulting in form list $\hat{l}$.

- The alignment between the current alignment A and $\hat{l}$ is performed using least distances (Section 13.1):

  - the distance between a form in $\hat{l}$ and a form-tuple in A can be calculated in different ways, for example, as the average of the distances from the form to each of the forms in the form-tuple in A;

  - forms from $\hat{l}$ that are matched to a form-tuple in A are added to the matched form-tuple;

  - forms from $\hat{l}$ that are not matched to a form-tuple in A are inserted as a new form-tuple in A.

This method thus allows forms that are not in the reference to be included in the alignment. But the overall alignment depends on which run is chosen as the reference, and the order in which the other runs are aligned to the running multiple alignment. This is because the matching depends on which forms that are already in the alignment.

13.11   Complete pair-wise alignment

All pair-wise alignments are calculated, thus $n(n-1)/2$ pair-wise alignments are performed. This method is not dependent on a reference run. However, it is more computationally expensive to perform.

The output from a pair-wise alignment is a list of matched forms, and the next step is to combine all the lists into an alignment containing form-tuples. The main challenge in this task is that there may be inconsistencies in the original lists, that is, that there may be a chain of matching peptides containing two different peptides from the same run.

Example Consider three runs i, j, k, and let $f_{i,1}$ be a form from run i. Assume that we find the following set of matches:

$$(f_{i,1}, f_{j,2})(f_{j,2}, f_{k,3})(f_{k,3}, f_{i,3}).$$

Then we will have an inconsistency if $f_{i,1}$ and $f_{i,3}$ are different.
$\triangle$

Such inconsistencies should not occur in the final alignment. This is a well known problem in informatics, and several solutions based on graph theory have been proposed. Often the strength (expressed as the similarity score) of the connection between paired forms is used to resolve the inconsistencies by removing some of the weaker pair-wise matches. A simple solution is to construct the alignment iteratively by picking the paired forms by decreasing strength (increasing distance). Pairs inconsistent with the current alignment are simply discarded.

Note that the final alignment can contain form-tuples that do not have forms from all samples (thus introducing missing values).

13.12   Hierarchical progressive alignment

Hierarchical progressive alignment means that subalignments (alignments of $\leq n$ runs) are progressively formed by pair-wise alignment of smaller subalignments until a single final alignment of all the runs is achieved.

The procedure consists of two main tasks:

- Determine the topology of the alignment, that is, a *guide tree* showing which subalignments are to be aligned in each iteration.

- Perform an alignment of two subalignments.

These subalignments can be performed successively or interwoven. Successive alignment means that the guide tree is constructed before the alignment, called a *static guide tree*. Interwoven means that the tree is constructed along with the aligning process; a so-called *dynamic guide tree*.

The distances (or similarities) between the subalignments are commonly used to determine which two subalignments should be combined in the next iteration. All the distances are stored in a two-dimensional table keeping track of the distances between the open subalignments, with 'open' here meaning that the subalignment is not part of any larger subalignment. Distances between subalignments are based on distances between runs (form lists). We therefore first discuss this problem.

13.12.1   Measuring the similarity or the distance of two runs

A simple way of measuring the similarity of two form lists is to let

- n_1, n_2 be the number of forms in the two runs respectively;

- n_c be the number of matching forms between the two runs.

The matching score is then:

$$S_M = \frac{2n_c}{n_1 + n_2}.$$

Analogously, the distance can then be defined as $1 - S_M$. Note that this score does not use the similarity between the matching forms.

In Mueller et al. (2007) S_M is combined with a score S_I calculated from how well the abundances of the matched forms correlate. This is calculated through the Spearman correlation coefficient, see Section 20.8.2. The final similarity measure between two runs is then defined as

$$S = S_M S_I.$$

Using abundance correlation is however only suitable when all the runs are from one situation, given that the abundances from different situations can of course vary (because they represent differentially abundant proteins).

13.12.2   Constructing static guide trees

The construction of a static guide tree starts by performing pair-wise alignments between every pair of form lists, and for each pair measuring the distance between the form lists of the pair. These distances are stored in the two-dimensional distance table D.

A node is constructed for each form list, and these nodes constitute the terminal nodes in the guide tree. A new inner node I is constructed with the nodes U, V

corresponding to the smallest value in D, as child nodes. Then the distances between I and all nodes O corresponding to the remaining open subalignments (i.e., the remaining runs in the first iteration) must be calculated and put into D. The smallest value in D then indicates which subalignments to align next.

The distances can be calculated as explained in Sections 21.1.5 and 21.1.9. Note that the distance calculations are based on the static pair-wise distances between the original form lists, and no alignment of subalignments is yet performed.

13.12.3 Constructing dynamic guide trees

When the construction of the tree and the alignment happens interweavingly, the subalignments of the nodes I and O are known, and the distance can be calculated from these two subalignments. One option is to use a generalization of S_M in Section 13.1. In addition to using the number of form-tuples one can also use the number of runs in each form.

13.12.4 Aligning subalignments

Aligning subalignments consists of retention time correction between the alignments, and then the matching of form-tuples. One procedure for the whole progressive alignment is described below.

Let the first form lists to be aligned be l_1 and l_2. Retention time correction is performed by changing l_1 to $\hat{l}_1$, thus achieving a subalignment of $(\hat{l}_1, l_2)$. Suppose that we construct the subalignment $(\hat{l}_3, l_4)$ further on in the procedure, and later decide to align these two subalignments. Then we can perform retention time correction between l_2 and l_4, resulting in $\hat{l}_2$. Change the retention times in $\hat{l}_1$ according to the changes in l_2 to achieve $\tilde{l}_1$, and then form the alignment. Note that the retention time of l_4 is unchanged in this process. Progressing in this way will ensure that a subalignment always includes a form list with unchanged retention times. This in turn means that an alignment of two such subalignments will always contain two original lists of retention times (one per subalignment) that can be used for retention time corrections.

To match the two form-tuple lists, we have to define a way of measuring the distance between two form-tuples, and the matching can then be done in the same way as when matching form lists.

13.12.5 SuperHirn

SuperHirn, Mueller et al. (2007) is an example of a tool using the hierarchical progressive approach. Here the identification is performed before the quantification, hence allowing the sequence of the peptide to be included as a feature in the form and used in the matching. The theoretical monoisotopic mass of the sequence is also a feature rather than the experimental monoisotopic mass, in order to avoid mass errors derived from the instrument.

13.13   Simultaneous iterative alignment

These methods generally start with an alignment of all the runs but taking into account only some of the forms. The alignment is then changed and increased by including more forms in an iterative way until final alignment is achieved.

As an example of simultaneous alignment we will again use XCMS. Two main tasks are explained: (i) how to construct the first (initial) alignment, and (ii) how to change and extend the alignment in each iteration.

13.13.1   Constructing the initial alignment in XCMS

The input for each run is a set of forms characterized by m/z, retention time, and abundance.

An interval length in the m/z dimension is defined such that all forms corresponding to the same peptide across the runs are assumed to have masses inside this interval. The length of the interval must therefore be determined in accordance with the accuracy of the instruments, for example, 0.25 m/z units, and the m/z range is then binned according to this accuracy. To avoid splitting a group of corresponding forms due to arbitrary bin boundaries, half-overlapping bins are used. An XIC is then constructed for each bin, using the peak intensities as calculated in Section 13.1. This is illustrated in Figure 13.7.

Determining initial form-tuples

To determine initial form-tuples the following procedure is performed for each mass interval (in the XIC):

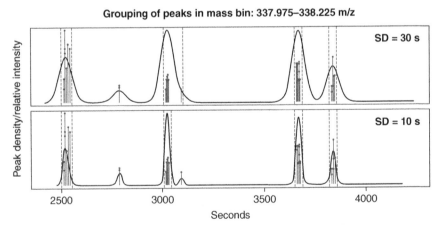

Figure 13.7   Construction of the initial alignment in XCMS. Individual forms are shown as sticks. Retention time is on the horizontal axis and abundance on the vertical axis. Reprinted from Smith et al. © 2006 by permission of American Chemical Society Publications.

1. Calculate the overall distribution of the forms. This is performed by using a Gaussian function for smoothing.[4] A continuous distribution is thus achieved. For clarification, a (chromatographic) peak of the distribution is called a *meta peak*. A meta peak is supposed to include a form-tuple. This is shown in Figure 13.7, where one can also see that the distribution depends on the standard deviation of the Gaussian function.

2. For each meta peak, in decreasing order:

   • descend down either side until the distribution increases again;

   • store all the forms in the meta peak in one form-tuple.

3. Prune or change form-tuples in relation to different conditions, for example:

   • prune tuples with forms from a low number of runs;

   • if the runs are from samples from different situations, all situations should have at least half the samples represented in the tuple;

   • for tuples containing several forms from the same run only one of them should be retained, either based on retention time or abundance.

Retention time corrections

The retention times can be aligned simultaneously in a single step using the initial form-tuples, and the procedure is as follows:

1. Select 'well behaved' form-tuples as a temporary standards. 'Well behaved' here means that almost all the runs have a form included, and that very few runs have more than one form included.

2. For each temporary standard:

   • calculate the median retention time;

   • calculate the deviation from the median for each form in the standard.

3. For each sample use a local regression fitting method to perform retention time correction. During this procedure outliers can be removed.

13.13.2   Changing the initial alignment

After retention time corrections, new form-tuples can be discovered, while existing ones can be changed or removed. The procedure above can therefore be repeated. One iteration is usually enough to reach a good convergence.

---

[4] Implemented by a kernel density estimation.

13.14   The end result and further analysis

The end result of a multiple alignment procedure is a list of form-tuples, for which there can be quite a few with many missing forms. Some of these missing values can be found by going back to the raw data spectra and searching for peaks where they ought to occur, based on the known forms in other runs.

The form-tuple list is in principle a two-dimensional table, with the forms as one dimension and the samples (runs) as the other. Abundance normalization should be performed for each run, and reasonable statistical analysis can be performed to find forms that seem to be differentially abundant across the situations. Other types of analysis can be performed if desirable, for example, clustering and discriminant analyses as described in Chapter 21.

13.15   Exercises

1. We have the following extracts from five succeeding raw data spectra:[5]

Mass		Intensities			
. . .		. . .			
334.5746	32148.21	69509.03	61643.49	50412.70	45717.94
334.5758	13601.00	74717.73	73270.56	66153.02	31619.20
334.5770	27915.30	63423.08	43250.24	53968.74	19406.67
334.5781	45610.90	38715.02	10244.35	25167.97	29063.98
334.5793	51255.68	20819.65	71951.88	21832.86	61498.03
334.5804	58113.01	27706.44	111507.45	51358.76	109100.73
334.5816	66092.03	31597.66	115819.74	74412.30	136306.40
334.5827	59962.06	23035.08	96170.92	81634.96	114528.03
334.5839	39776.81	10755.78	69033.30	70593.06	52742.18
334.5850	17996.37	18404.88	46532.11	55860.80	0.00
334.5862	3752.39	34213.64	34423.51	64986.09	28255.43
. . .		. . .			

Construct an XIC for the m/z interval [334.5750 - 334.5850] by using the largest intensity from each slice.

2. A form has the features m/z, retention time and abundance. We want to compare two forms with values (605.23, 34.3, 201) and (605.32, 37.9, 149). Calculate the distance between the forms using the three different measures below (see Section 13.9.1):

---

[5] Note that for simplicity of presentation the masses have been rounded to 4 decimals.

(a) the simple measure of 0/1 using the thresholds $(0.1, 3.0, 60)$ for the differences between the feature values;

(b) the Euclidean distance (Section 21.1.1) with the weights $(100, 0.1, 0.0003)$;

(c) the revised Mahalanobis distance with the user defined thresholds from (a).

3. We want to constructs alignments between two form lists $F = \{f_1 \cdots f_9\}$ and $G = \{g_1 \cdots g_8\}$. Assume that we have a distance threshold of 3 and found the following matches below the threshold, where the values are the distances between the forms:

$(f_1, g_2 : 2.4)(f_2, g_3 : 2.9)(f_3, g_4 : 2.8)(f_4, g_3 : 1.7)(f_4, g_4 : 2.3)(f_5, g_6 : 2.4)$
$(f_6, g_5 : 1.2)(f_7, g_7 : 0.9)(f_8, g_7 : 2.2)(f_9, g_8 : 1.3).$

(a) Find the alignment using the method with least distance. Assume that the requirement for significant difference between the best and second-best is satisfied.

(b) Find one of the alignment having the maximum number of matches.

4. We want to perform complete pair-wise alignment for three form lists F, G, H. From the three pair-wise alignments we have found the following matches:

$(F, G) : \cdots (f_6, g_4 : 2.0)(f_7, g_5 : 2.8) \cdots$
$(G, H) : \cdots (g_4, h_4 : 3.1)(g_5, h_5 : 2.4) \cdots$
$(F, H) : \cdots (f_6, h_5 : 1.9) \cdots$

(a) Explain the inconsistency.

(b) Discuss different ways of resolving the inconsistency.

(c) We have assumed that the pair-wise alignments are unambiguous. Would it be simpler to resolve inconsistencies if ambiguity in the pair-wise alignments was allowed?

5. For six runs we have calculated the pair-wise similarity as follows:

	R2	R3	R4	R5	R6
R1	7.3	4.6	9.8	2.8	6.8
R2		10.5	7.2	5.0	2.7
R3			4.5	5.3	9.0
R4				1.2	7.3
R5					4.3

(a) Which runs will be used in the first subalignment?

(b) Assume that the similarity between two subalignments A, B (a single run is considered as a subalignment) is calculated as:

$$\frac{1}{nm} \sum_{i=1}^{n} \sum_{j=1}^{m} s(a_i, b_j),$$

where

- n, m are the numbers of runs in A, B respectively;

- a_i and b_j are runs in A, B respectively;

- $s(a_i, b_j)$ is the similarity between the two runs.

Calculate the new similarity table, and determine the next subalignments to align.

13.16   Bibliographic notes

Retention time correction   Silva et al. (2005), Smith et al. (2006), Mueller et al. (2007), Podwojski et al. (2009), Vandenbogaert et al. (2008).

Pair-wise matching   Prince and Marcotte (2006), Lange et al. (2007), Mueller et al. (2007).

Reference based alignment   Higgs et al. (2005), Lange et al. (2007).

Complete pair-wise alignment   Li et al. (2003), Li et al. (2005) SpecArray.

Hierarchical progressive alignment   Mueller et al. (2007) SuperHirn.

Simultaneous iterative alignment   Smith et al. (2006), Tautenhahn et al. (2010) XCMS, Voss et al. (2011).

Review of label free quantitative mass spectrometry   Neilson et al. (2011).

14

Label free quantification by MS/MS spectra

In the preceding chapter we observed that performing label free quantification based on ion currents entailed several complicated calculations. Studies have therefore been carried out to see whether equally good results can be achieved through simpler methods. In this chapter we will introduce label free methods where MS/MS spectra are used for both the quantification and the identification.

14.1   Abundance measurements

The basic idea in using MS/MS spectra for quantification is that there ought to be a correlation between the number of times peptides from a given protein are selected for MS/MS scans, and the abundance of the protein in the sample.

When considering such methods there are several points to keep in mind:

- Larger proteins result in more peptides than smaller proteins, and therefore yield more expected MS/MS scans.

- Different peptides have different *a priori* probabilities for being selected for MS/MS analysis, for example, the ionization of a peptide depends on its amino acid composition.

- The results depend on the quality of the peptide identifications, since false positive identifications can add substantial noise to the calculations.

- An identified peptide can belong to several proteins, complicating the analysis.

- The statistical analysis becomes problematic when only a low number of peptides are identified for a protein (1 to 3 peptides).

Computational and Statistical Methods for Protein Quantification by Mass Spectrometry,
First Edition. Ingvar Eidhammer, Harald Barsnes, Geir Egil Eide and Lennart Martens.
© 2013 John Wiley & Sons, Ltd. Published 2013 by John Wiley & Sons, Ltd.

- Counting algorithms yield sharply discrete value distributions (1, 2, 3, 4, etc.) whereas intensity-based algorithms tend to yield more continuous results.

The approach described in this chapter can be used both for comparing the abundances of a protein across two or more samples, and for comparing the abundances of two or more proteins in a single sample. Although we are focusing on the first approach, we will also mention methods mainly using the second approach. However, especially due to different ionization probabilities for different peptides, the use of the approach for comparing different proteins is generally less reliable.

Abundance measurements can be based on different types of observations:

Sequence coverage: The fraction of the protein sequence that is covered by matched MS/MS spectra.

Peptide count: The number of peptides in the protein that are matched by MS/MS spectra.

Spectral count: The number of MS/MS spectra that are matched to peptides in the protein.

Example   Say that we have the protein sequence below, where tryptic *in silico* digested peptides are shown. Note that we assume that there are no missed cleavages (except when there are two consecutive arginines or lysines).

```
MALLPR | ALSAGAGPSWR | RAAR | AFR | GFLLLLPEPAALTR |
ALSR | AMACR | QEPQPQGPPPAAGAVASYDYLVIGGGSGGLASARR |
AAELGAR | AAVVESHK | LGGTCVNVGCVPK | VMWNTAVHSEFMHD-
HADYGFPSCEGK | FNWR | VIKEK | RDAYVSR | LNAIYQNNLTK |
SHIEIIR | GHAAFTSDPKPTIEVSGK | KYIAVEMAGILSALGSK |
TSLMIR | HDKVLR
```

Assume that the following peptides are matched by the indicated number of MS/MS spectra:

```
ALSAGAGPSWR: 2
AAELGAR: 1
LGGTCVNVGCVPK: 3
VIKEK: 1
SHIEIIR: 3
KYIAVEMAGILSALGSK: 2
```

The abundances measured by the different methods are then:

- sequence coverage: 27.7 % (60 out of 217 residues);

- peptide counts: 6;

- spectral counts: 12.

△

For the peptide counts there are in addition several variants; the same peptide can be counted several times depending on charge state, modification state, and missed cleavage sites.

The problem of shared peptides discussed in Chapter 5.9 must also be taken into account. As usual the straightforward approach is to include only uniquely identified peptides.

Some additional issues to keep in mind are:

- Spectral counting assumes a linear relationship between the number of MS/MS spectra observed for a protein and its relative abundance. Such linearity has been observed to hold over two orders of magnitude.

- Spectral counting strongly depends on peptide identification. If the number of identified peptides is too low, it is very difficult to perform proper statistical testing.

- Uncertainties in the protein identifications must be treated properly.

14.2 Normalization

There are different ways of normalizing the measured values:

- normalize each protein in a run (sample), for example, dividing by the length of the protein;

- normalize across the runs – global normalization, for example, dividing by the sum of the measured values.

14.3 Proposed methods

Several methods have been proposed for MS/MS based quantification using peptide identifications. We will here mention some of the most popular approaches, highlighting the variation in how the abundances can be determined. Note that some of the methods include normalization, while others do not. Furthermore, some calculate an expression for the abundances in only a single sample, and then leave it to the user to calculate relative abundances for proteins across two samples. Others directly calculate relative abundances, providing the ratio of protein abundances between two samples.

References for the various measurements are given in the Bibliographic notes. In all formulas we use P for the protein under consideration.

14.4 Methods for single abundance calculation

Note that 'single' is here used as opposite to relative, as introduced in Chapter 5.8.

14.4.1 emPAI

emPAI (exponentially modified Protein Abundance Index) is based on sequence coverage.

$$emPAI_P = 10^{PAI_P} - 1; PAI_P = \frac{n_i}{n_o},$$

where:

- n_i is the number of identified peptides for the protein;

- n_o is the number of identifiable peptides in the protein.

Note that emPAI uses peptide counts, but remember that peptide counts can be calculated in different ways. In the original article for PAI a peptide occurring with different charge states and/or modification states is counted separately for each charge and each modification state, and also for each missed cleavage site. As a result, PAI can be larger than one.

Also note that an extensive determination of n_o for a protein is not straightforward, given that the probability for a peptide to be identified depends both on its amino acid composition as well as on the background of peptides in which it elutes. However, in most implementations it is only the peptide mass that is used to constrain identifiable peptides, with peptides falling within a given mass range considered identifiable, whereas those outside of this mass interval are considered nonidentifiable.

14.4.2 PMSS

Peptide Match Score Summation (PMSS) uses the scores of the peptide identifications to estimate protein abundance. With this method 2.5–5 fold changes are reported with 90–95% confidence. The score E_P for a protein P is:

$$E_P = \sum_{p \in T_P} \sum_{\sigma \in I_p} s(\sigma, p),$$

where:

- T_P is the set of tryptic peptides in P;

- p is a tryptic peptide;

- I_p is the set of experimental spectra identifying p;

- $s(\sigma, p)$ is a normalized score of the match between a spectrum σ and p.

Note that the double summation is used because one protein can have many different identified tryptic peptides (outer summation), and each tryptic peptide can be identified by many spectra (inner summation).

14.4.3 NSAF

The Normalized Spectral Abundance Factor (NSAF) for a protein P is defined as:

$$NSAF_P = \frac{s_P/l_P}{\sum_{i=1}^{n} s_i/l_i},$$

where:

- s_x is the spectral count for protein x;
- l_x is the length of protein x;
- n is the number of proteins in the database.

Note that both the length of the protein as well as the other proteins are included in this measurement.

14.4.4 SI

Spectral Index (SI) uses the total intensities of the identified MS/MS spectra:

$$SI_P = \sum_{p=1}^{m} \left(\sum_{j=1}^{c_p} I_j \right),$$

where:

- m is the number of peptides identified for P;
- c_p is the spectral count for peptide p;
- j is the jth spectrum of the c_p spectra;
- I_j is the total intensity of spectrum j.

SI thus inherently incorporates both spectral counts as well as MS/MS intensities. A normalized version of SI, SIN is defined as:

$$SIN_P = \frac{SI_P}{\sum_{i=1}^{n} SI_i} \frac{1}{L_P}, \tag{14.1}$$

where:

- n is the number of quantified proteins;
- SI_x is SI for quantified protein x;
- L_P is the length of P.

14.5   Methods for relative abundance calculation

These methods calculate a ratio for each protein from the single abundances in two samples.

14.5.1   PASC

Protein Ratios from Spectral Counts (PASC) includes a normalization for measuring the ratio of a protein's abundance in two samples:

$$PASC_P = \log_2 \left[\frac{\frac{c_1+f}{t_1-c_1+f}}{\frac{c_2+f}{t_2-c_2+f}} \right] = \log_2 \left[\frac{c_1+f}{c_2+f} \right] + \log_2 \left[\frac{t_2-c_2+f}{t_1-c_1+f} \right],$$

where:

- c_1, c_2 are the spectral counts of the protein in the two samples, respectively;

- t_1, t_2 are the total spectral counts over all proteins in the two samples;

- f is a correction factor (0.5), included to avoid discontinuity if one of the spectral counts is zero.

The measurement is inspired by Serial Analysis of Gene Expression (SAGE).

14.5.2   RIBAR

One essential property of RIBAR (Robust Intensity Based Averaged Ratio) is that *only peptides occurring in both samples are used* in the quantification of each protein. This restriction is used to ensure consistent and reproducible quantification across many replicate analyses, a feature missing from other MS/MS based label free methods, as the run-to-run relative standard deviation is typically very large for these other methods (on the order of 100 %).

A peptide can be identified in several modification states (e.g., amino acids can be acetylated, oxidized, etc.) including a separate state for unmodified peptides. We refer to a peptide and its modification state as a *peptidemod*. Thus a peptide can exist as several peptidemods.

To explain RIBAR we define:

- Q is the set of peptides uniquely identifying P that are observed in both samples;

- M is the set of peptidemods associated with Q, note that a peptidemod must also be identified in both samples;

- $S_{m,1}, S_{m,2}$ are the sets of MS/MS spectra uniquely identifying peptidemod m in the two samples respectively;

- I_s is the total intensity of an MS/MS-spectrum s.

Define the relative abundance for a peptidemod m as:

$$r_m = \log_2 \frac{\sum_{s \in S_{m.1}} I_s}{\sum_{s \in S_{m.2}} I_s},$$

and the relative protein abundance is calculated as

$$R_P = \frac{\sum_{m \in M} r_m}{|M|},$$

with $|M|$ the number of elements in M, thus obtaining an average across the relative peptidemod abundances.

14.5.3 xRIBAR

RIBAR relies only on peptides found in both samples and therefore cannot calculate ratios for proteins that do not have peptides occurring in both samples. An extended version of RIBAR, called xRIBAR has therefore been proposed. Here, an extended ratio r_x is calculated as:

$$r_x = \log_2 \frac{\frac{\sum_{s \in S_1} I_s}{n_1}}{\frac{\sum_{s \in S_2} I_s}{n_2}},$$

where:

- S_1, S_2 are the set of MS/MS spectra that uniquely identify P in the indicated sample, but not the other;
- n_1, n_2 are the number of spectra in S_1, S_2 respectively.

This additional ratio r_x is added to M, creating M'.
xRIBAR is then calculated as:

$$R'_P = \frac{\sum_{m' \in M'} r_{m'}}{|M'|}.$$

Note that $|M'| = |M| + 1$. All spectra identifying peptides that are not shared between samples thus collectively contribute only one additional, averaged measurement to the overall protein ratio calculation, thus having a relatively small impact on proteins that also have one or more peptides that are shared between samples. If a protein has no common peptides between the two samples, the extended ratio r_x will be the only element in M'.

Because of the inclusion of peptides that are not shared between samples, xRIBAR yields a slightly higher relative standard deviation than RIBAR across replicate runs. However, even this increased relative standard deviation remains well below that of other methods discussed above. The choice between RIBAR and xRIBAR thus depends on the reproducibility of the run, the desired level of precision of the measurements across replicates, and the desired coverage of quantifiable proteins across replicates. In practice, it is straightforward to calculate both RIBAR and xRIBAR at the same time since the calculations are largely parallel, so both metrics can easily be provided for evaluation.

14.6   Comparing methods

A measure used for protein abundance determination ought to satisfy the following requirements:

1. have a linear correlation with protein abundance;

2. have high precision;

3. have high accuracy.

Consider a protein in two samples to be compared, with relative abundances to the total sample abundances r_1 and r_2 respectively, and the measured values (sequence coverage, peptide counts, or spectral counts) being c_1 and c_2. The first requirement then means that $\frac{c_1}{c_2} = \frac{r_1}{r_2}$ (using normalization to total abundance).

Liu et al. (2004) found that of the three options – sequence coverage, peptide counts, and spectral counts – spectral counts best fit the first two requirements, having both high reproducibility and a linear dynamic range over 2 orders of magnitude.

Below we describe two recently performed comparison studies.

14.6.1   An analysis by Griffin

Griffin et al. (2010) performed an extensive comparison of some of the methods.

Normalization

Different ways of normalizing SI was analyzed by considering the differences between the results from a set of technical replicates, and it was concluded that SIN was superior to the others, as it showed no significant differences between the results from the replicates.

They also replaced SI with SC (spectral count), and SI by PN (peptide count) in Equation 14.1, but this resulted in significant differences between the replicates. They also tested $PASC$ and $NSAF$, but the conclusion was that SIN was better in reducing the variability. Substituting SI for SC in $NSAF$ made $NSAF$ better, but they still observed significantly different results.

Abundance measurements

In a complex protein mixture containing 19 protein standards over a dynamic range of 0.5–50 000 fmol they got a correlation coefficient $R^2 = 0.9239$ between the

SIN measurement and the actual protein load. The slope of the regression line was 1.223.

They also compared SIN to SC using the area under the curve (AUC) to see how well either predicts the abundance of proteins in a standard protein mixture, and concluded that SIN was superior.

14.6.2   An analysis by Colaert

Colaert et al. (2011c) compared emPAI, NSAF, and SIN for precision and accuracy. Their results indicate that the spectral counting method narrowly outperforms the protein sequence coverage-based emPAI, with the MS/MS spectral intensity-based SIN method showing better accuracy but worse precision.

In Colaert et al. (2011a) RIBAR and xRIBAR were also compared to the three methods mentioned above, showing similar performance to NSAF and emPAI within a single run, but with much higher across-replicate reproducibility. Interestingly, in this study the SIN measurement performed worse in terms of reproducibility than the spectral counts or sequence coverage measurements, whereas the study by Griffin discussed above came to the opposite conclusion. However, since RIBAR outperforms all other algorithms in terms of reproducibility by simply focusing on shared peptides, it may well be that the dataset by Griffin contained quite a few more shared peptides across replicates than the two standard datasets used by Colaert, allowing the SIN metric to perform well in their specific case. Indeed, the RIBAR and xRIBAR measurements are not that different from SIN, as they both rely on spectral intensity.

Because the Colaert study used two independent, complex, and standardized data sets that closely resemble realistic samples, it seems reasonable to maintain the constraints imposed by RIBAR when applying an MS/MS based label free quantification approach.

14.7   Improving the reliability of spectral count quantification

Although studies have shown good correlation between spectral counts and actual protein abundances, the correlation is still poor when low spectral counts are observed for a protein. Several techniques are proposed to improve the reliability for low abundance proteins.

Zhang et al. (2009) investigated how the number of spectral counts depends on the dynamic exclusion time (DE), that is, the number of seconds that a chosen mass is excluded from being selected again for MS/MS processing. As expected, the number of qualified spectra decreases (exponentially) with increasing DE. However, the number of identified peptides and inferred proteins first increased and then decreased with a maximum at a DE of 90 seconds. Due to these seemingly conflicting results they investigated how the quantitative results varied with the DE, using $NSAF$. They found that the optimal dynamic exclusion time is proportional to the average chromatographic peak width at the base.

Zhou et al. (2010) described a method where low scoring peptide identifications are also taken into account. In the first step only the significant identifications are counted. Then in a second step the low scoring identifications (peptides) for which high scoring identifications also exist are considered as well. Based on several properties they validated the considered identifications, and in that way increased the number of spectral counts, observing an increase of more than 20 %. The validation was based on parent ion mass error distribution, retention time distribution, and a comparison of the low and high scoring spectra of the same peptide.

A variant of the technique above is to compare the nonidentified spectra to the identified spectra, for example, by a clustering procedure. A nonidentified spectrum that is sufficiently similar to an identified spectrum is counted.

14.8   Handling shared peptides

Zhang et al. (2010) investigated the possibility of distributing spectral counts for a shared peptide over the proteins containing the peptide. They used NSAF and analyzed several ways of distributing the spectral counts by comparing the results to the known protein abundances in samples.

To illustrate this study we consider a set of proteins and a set of identified peptides. A general form of $NSAF$ for protein P is defined as:

$$NSAF_P^* = \frac{C_P + \sum_{i \in S_P} c_i d(Q_i)}{l_1(U_P) + l_2(S_P)},$$

where:

- U_P is the set of unique peptides in P;

- S_P is the set of shared peptides in P;

- c_i is the spectral count for peptide i;

- Q_i is the set of proteins containing peptide i;

- $C_P = \sum_{i \in U_P} c_i$ is the spectral count sum of all the unique peptides in protein P;

- d is a distributing function;

- l_1, l_2 are functions for calculating lengths depending on the unique and shared peptides respectively.

They investigated different forms of the three functions d, l_1, l_2, and found that distributing the spectral counts of a shared peptide proportional to the unique spectral counts resulted in a linear relation between protein abundance and NSAF over a dynamic range of at least three orders or magnitude. The obtained general form $NSAF_P^*$ is:

$$NSAF_P^* = \frac{C_P + \sum_{i \in S_P} c_i \frac{C_P}{\sum_{j \in Q_i} C_j}}{L_P},$$

where L_P is a length of P, calculated such that it can be the whole protein length or a length depending on the part of the protein covered by the unique and shared peptides.

Example We have a group of three proteins, each containing the identified peptides:

- $P_1 : 1, 2, 3, 4, 5$;

- $P_2 : 2, 3, 6, 7, 8$;

- $P_3 : 3, 5, 7, 9$.

Thus the unique peptides of P_1 are peptides 1 and 4, and

$$NSAF_{P_1}^* = \frac{c_1 + c_4}{L_{P_1}} \left(1 + \frac{c_2}{c_1 + c_4 + c_6 + c_8} \right.$$

$$+ \left. \frac{c_3}{c_1 + c_4 + c_6 + c_8 + c_9} + \frac{c_5}{c_1 + c_4 + c_9} \right).$$

$\triangle$

14.9   Statistical analysis

The general methods for statistical analysis described earlier can be used. More specific methods for spectral counts are found in the Bibliographic notes below.

14.10   Exercises

1. We consider a protein P for which the theoretical tryptic peptides are shown in Table 14.1. For each peptide the spectral counts in two LC-MS/MS runs are given.

Table 14.1   The spectral counts for each peptide.

Peptides	SC_1	SC_2	Peptides	SC_1	SC_2	Peptides	SC_1	SC_2
MALLPR	0	0	ALSAGPSWK	6	4	QAAAFR	0	1
AAGASYDK	12	6	GSGGLASAK	0	0	AAELGAR	8	4
GFLLLPEPR	0	0	ALSAMACK	11	4	QEPQQGPP	0	0

(a) Calculate $emPAI_P$ for run 1, assuming that all peptides are identifiable.

(b) How many score values will have to be added to be able to use PMSS for run 1?

(c) Calculate $NSAF_P$ for run 1. Assume (for simplicity) that there are six additionally identified proteins in the sample, and that the spectral counts and lengths of these are: (13, 50) (2, 80) (64, 110) (27, 98) (14, 110) (30, 78).

(d) Assume that the intensities of the spectra used for identifying the peptides in run 2 are: (Peptide 2: 930.24, 1120.56, 890.45, 1845.45, 1212.46, 987.38) (Peptide 4: 1212.46, 987.38, 2409.45, 1745.34, 1146.34, 904.56) (Peptide 6: 1620.45, 1934.56, 998.45, 1734.45) (Peptide 8: 986.45, 1434.56, 1924.67, 1645.69). Calculate SI_P for run 2.

(e) Assume that the total spectral counts for the two samples are 4875 and 4591. Calculate $PASC_P$.

(f) Assume that each identified peptide in Table 14.1 is either identified as unmodified or carrying a specific modification, and that the spectral counts for these are:

Peptides	Sample 1		Sample 2	
	Unmodified	Modified	Unmodified	Modified
ALSAGPSWK	4	2	4	0
QAAAFR	0	0	0	1
AAGASYDK	9	3	6	0
AAELGAR	4	4	3	1
ALSAMACK	11	0	4	0

How many spectra will be used to calculate the relative abundance for the precursor protein using RIBAR?

14.11   Bibliographic notes

PAI, emPAI   Rappsilber et al. (2002), Ishiama et al. (2005a).
PMSS   Allet et al. (2004), Colinge et al. (2005).

NSAF   Pavelka et al. (2008).

SI   Griffin et al. (2010).

PASC   Old et al. (2005), Griffin et al. (2010).

RIBAR, xRIBAR   Colaert et al. (2011c), Colaert et al. (2011a).

Improving the reliability   Zhang et al. (2009), Zhou et al. (2010), Carvalho et al. (2009), Dicker et al. (2010).

Shared peptides   Zhang et al. (2010), Zybailov et al. (2006), Jin et al. (2008).

Statistical analysis   Zhang et al. (2006), Choi et al. (2008), Cooper et al. (2010), Booth et al. (2011).

15

Targeted quantification – Selected Reaction Monitoring

All methods discussed so far have been discovery oriented quantification methods, where in principle one tries to quantify all proteins in the sample(s). However, when one *a priori* knows which proteins to quantify, other methods are more appropriate, and these are called targeted quantification methods.

In this chapter we will describe the most common targeted quantification method, called Selected Reaction Monitoring (SRM).[1]

15.1 Selected Reaction Monitoring – the concept

Mass spectrometry generally produces two levels of spectra: MS spectra and MS/MS spectra. A peptide can be identified by observing a peptide precursor mass-over-charge m/z or mass m in an MS spectrum and a set of fragment ion masses $f_1, f_2, \ldots, f_n$ in the corresponding MS/MS spectrum.

The idea behind SRM is as follows: *If for a protein P there exists a peptide precursor mass m and a fragment ion mass f that together uniquely determine P in the sample under consideration, then one can target only this pair in the mass spectrometer. If a signal is found for this selected pair, the protein can be inferred to be in the sample, and the intensity of the signal can be used for quantification of that protein.*

---

[1] Note that SRM is also sometimes referred to as Multiple Reaction Monitoring (MRM). The International Union of Pure and Applied Chemistry (IUPAC) has however defined SRM to be the correct term and we will therefore exclusively use SRM.

Computational and Statistical Methods for Protein Quantification by Mass Spectrometry, First Edition. Ingvar Eidhammer, Harald Barsnes, Geir Egil Eide and Lennart Martens. © 2013 John Wiley & Sons, Ltd. Published 2013 by John Wiley & Sons, Ltd.

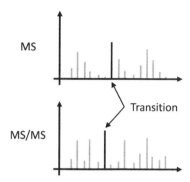

Figure 15.1   A peptide precursor and fragment ion mass pair is referred to as a transition. An example transition is indicated. Note that selecting the same MS peak but a different MS/MS peak results in a different transition. A single precursor can thus yield multiple transitions.

A single peptide precursor and fragment ion mass pair, (m, f), is referred to as a *transition*, as shown in Figure 15.1, and transitions are the key element in SRM.

Given that only a small subset of the peptides present in a sample are quantified in SRM, the selection of which peptides to use for a particular protein is of the utmost importance. This is covered in detail in Section 15.6. First, however, we describe the specific mass spectrometer type commonly used for SRM, and how this instrument is used in the SRM runs.

15.2   A suitable instrument

SRM requires the targeting of specific masses at both the MS and MS/MS level, requiring special instruments. Most commonly triple-quadrupole instruments, often simply called *triple-quads*, are used for this purpose, as shown in Figure 15.2.

As the name implies, a triple-quad consists of three quadrupoles, each with a distinct role in the SRM analysis:

- Q1: Functions as a peptide mass (actually m/z) filter;
- Q2: Is used to fragment the selected peptide(s);
- Q3: Functions as a fragment ion mass (actually m/z) filter.

This setup makes it possible to target specific transitions, and in the process filter out noninteresting peptides and corresponding fragment ions.

It is however also possible to use an iontrap or QqTOF to produce so-called *pseudo-SRM* spectra. The difference with the triple-quads is that the fragment ion selection is not performed physically, as in the third quadrupole of the triple-quad, but electronically, by limiting the scanning range of the fragment ion mass analyzer (iontrap or TOF for an iontrap or QqTOF instrument, respectively). This allows these

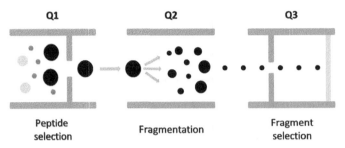

Figure 15.2   Triple-quad instrument in SRM mode: The first quadrupole (Q1) performs peptide filtering, the second quadrupole (Q2) fragments the selected peptide(s), and the third quadrupole (Q3) filters based on fragment ion masses. Overall, the effect is that only targeted transition(s) hit the detector indicated as a light grey bar on the far right.

instruments to ignore the majority of the m/z range of the MS/MS spectrum, and thus to spend more time analyzing a dedicated m/z window, centered on the fragment ion of interest. In general however, triple-quads currently remain the preferred option for SRM analyses.

15.3   The LC-MS/MS run

Remember that the order in which the peptides enter the MS instrument depends on the physiochemical properties of the peptides, that is, peptides with similar properties elute from the LC in roughly the same time interval.

As soon as a peptide elutes from the LC it enters the MS instrument and goes through the SRM triple-quad filtering process described above. The output from this process can be considered as a mass chromatogram, and can be separated based on the individual transitions.

Before starting the run, the selected transitions for the targeted proteins are first collected in a list, typically with 2–5 transitions for each selected peptide for redundancy and better statistical power. This list is then provided to the instrument, which upon starting the run will cycle through the listed transitions. The time interval required to perform one scan for all listed transition is referred to as the *cycle time*. The cycle time thus provides the amount of time that the instruments can take to process an entire list of transitions. Note that the cycle time is a user-specified parameter.

Also note that both the MS and the MS/MS part of the instrument are scanning in selected ion mode (described in Chapter 5.7), and a selected ion chromatogram (SIC) is made for each transition. Whenever a specific transition in the list is scanned, an intensity corresponding to the fragment intensity is added to its SIC. For example, if a peptide elutes from the LC in a 30 second interval and one is using a cycle time of 3 seconds, 10 peaks will be added to the transition's SIC.

The process is illustrated in Figure 15.3.

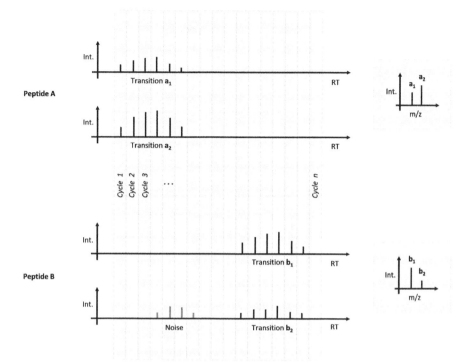

Figure 15.3   Illustration of the targeting of transitions. Two peptides, A and B, with two fragment ions per peptide, are targeted in an SRM run. The instrument cycles through the list of four transitions, and whenever a transition is scanned the detected intensity is added to the transition's selected ion chromatogram (SIC). In this example the transitions for peptide A are detected in Cycle 1–Cycle 6, while peptide B elutes later, shifting its transitions further along the LC gradient. A transition occurring at several locations, as shown for transition b_2, indicates noise. On the right, the detected transitions are plotted as spectra.

As the instrument cycles through the list of transitions it can spend a defined time, called the *dwell time*, on each transition. For example, if 10 peptides are monitored, with 5 fragment ions each (for a total of 50 transitions) and the cycle time is one second, a dwell time of 20 ms (1000 ms / 50) is possible. In other words, each transition on the list is monitored for 20 ms (dwell time), and it thus takes 1000 ms (1 second; cycle time) to record the full list of 50 transitions. Dwell time is illustrated in Figure 15.4.

Between each transition there is also a very short time interval called the *inter scan delay* or the *inter scan time*. When the dwell times and the inter scan times are very small, and the sample concentrations are high, a storage of product ions can take place in the collision cell. In other words, fragment ions from the previous transition can still be in the collision cell when the next transition is monitored. This can result in so-called *cross talk*, that is, the fragment ions from one transition are scanned out during another transition. Even a 0.01 % cross talk effect can result in

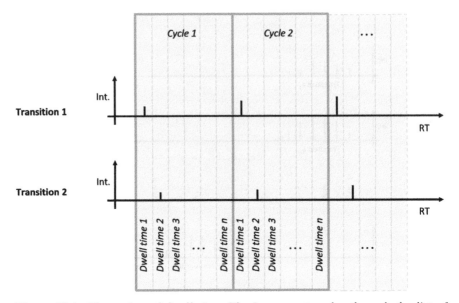

Figure 15.4   Illustration of dwell time. The instrument cycles through the list of transitions spending a defined time, called the dwell time, on each transition. Here two transitions in two cycles are shown. The first transition is detected in the first dwell time interval in each cycle, while the second transition is detected in the second dwell time interval in each cycle. Note that the sum of all dwell times in a cycle is equal to the cycle time.

false positives being observed in the resulting quantitative data. However, improved triple-quad instruments have drastically reduced, or even removed, the cross talk effect.[2]

Note that MS and MS/MS spectra are not constructed when performing SRM experiments. Instead only the SICs, produced on the MS/MS level are used for quantification. From each SIC an abundance value is calculated by taking the sum of the peaks in the SIC, as shown on the right hand side in Figure 15.3. The protein abundance can then be calculated from the abundances of all the transitions corresponding to the protein, as explained in Chapters 5.8 and 5.9. Abundances calculated from the SICs correspond to the elementary abundances discussed in Chapter 5.8.

Also note that identification is not necessary for SRM assays, but it can be performed for validation purposes, as will be explained in Section 15.7.

15.3.1   Sensitivity and accuracy

It is important to keep in mind that the number of transitions that can be quantified with high sensitivity and high accuracy in a single LC-MS/MS run is inherently

---

[2] Cross talk should not be confused with carryover in the HPLC module.

limited. This is due to the way in which the instrument processes the transitions as explained above. For sensitivity and accuracy we have:

- High sensitivity requires long enough dwell times to accumulate sufficient signal for detection in one measurement.

- High accuracy requires enough observations to recreate the SICs.

By increasing the cycle time, a larger number of transitions can be analyzed without having to lower the dwell time. But this will result in less frequent recording of any given transition (the transition is said to be 'active' less often), in turn reducing the accuracy of the measurement. Another way to increase the number of transitions that can be monitored is to reduce the dwell time, while keeping the cycle time constant. But this strategy has the disadvantage that the lower dwell time can be too short to accumulate sufficient signal for each transition, thus reducing the sensitivity.

The relationship between sensitivity and accuracy in terms of dwell time and cycle time is illustrated in Figure 15.5. The area in grey is the actual peak to be measured.

- From (a) and (b) we see that increasing the dwell time with constant cycle time increases the accuracy.

- From (a) and (b) we see that too short dwell times decreases the sensitivity, as most of the intensities at the edges of the peak are too low to be detected or distinguished from noise.

- From (c) and (d) we see that long dwell times yields very high accuracy for the individual data points, but low overall accuracy due to the limited number of data points.

Several approaches have been proposed to render SRM more efficient in terms of sensitivity:

- *Scheduled SRM.* A transition is only acquired during a limited time window around each peptide's (predicted) retention time. In this case the retention time must be known, either from previous experiments or from theoretical calculations. The transition list to monitor will thus change during the run, and the instrument must support such time-dependent transition lists.

- Optimize the dwell times depending on peptide abundance. Shorter dwell times are used for high abundance peptides, since these will still yield sufficient signal under these conditions. This requires that the abundances of the peptides are approximately known *a priori*, that the abundance limit for obtaining sufficient signal for a given dwell time can be determined, and that the instrument supports different dwell times for the different transitions on the list.

In general, the SRM setup provides high sensitivity, due to the fact that co-eluting background ions are filtered out during analysis, and because specific transitions are actively targeted while ignoring anything else. This particular focus on predefined targets thus allows SRM to achieve reliable quantification of low-abundance proteins in highly complex mixtures.

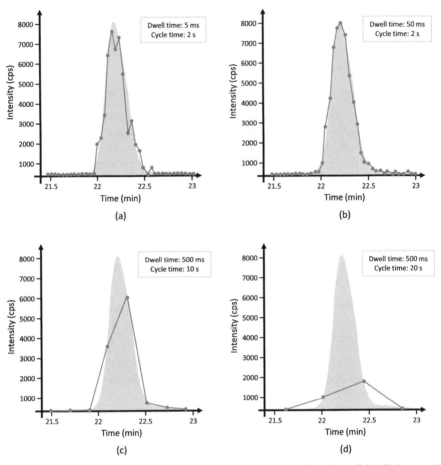

Figure 15.5   Illustration of accuracy and sensitivity as a function of dwell time and cycle time. The intensity is plotted in normalized form as 'counts per second.' The figure is based on a figure in Lange et al. (2008).

15.4   Label free and label based quantification

Similar to the division used for discovery-oriented quantification, SRM based quantification can be divided into label free and label based approaches. The label based approach increases the complexity and cost of an experiment as compared to the label free approach, but because of the higher precision and more accurate quantification, label based SRM is the most commonly used option.

15.4.1   Label free SRM based quantification

The simplest application of SRM is found in the label free quantification approach, where relative quantification is based on the signal intensities of specific SRM transitions.

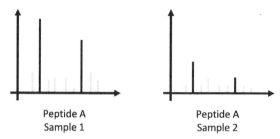

Peptide A
Sample 1

Peptide A
Sample 2

Figure 15.6   Label free quantification using SRM. The same peptide is found in two separate samples, here visualized with the normalized MS/MS spectra. Selected fragment ion peaks are shown in black, while the other peaks are shown in light grey as a reference (although these are not detected when performing an SRM experiment). By comparing the selected fragment ion peaks, it becomes clear that Sample 1 has a higher abundance of the given peptide, and correspondingly the protein, than Sample 2.

A challenge with this approach is the variation in signal intensity from one experiment to the next and even within one experiment, that is, a peptide yields different intensities depending on the sample and time of analysis. This issue of variation becomes increasingly important as more samples are analyzed. Normalization therefore has to be carried out, most often relying on the standard assumption that most protein/peptide abundances do not change between samples. The abundances of the individual transitions are then normalized to achieve equal total abundances in each experiment.

Figure 15.6 shows how label free SRM can be used for relative quantification. Keep in mind that label free quantification using SRM is affected by the typical issues for label free quantification as described in Chapter 13. However, label free quantification presents a simpler problem for targeted quantification than for discovery quantification in that the comprehensive determination of corresponding features (peaks) are avoided or reduced (the corresponding transitions are known).

15.4.2   Label based SRM based quantification

Label based quantification can be performed by labeling one of the samples, combining the samples and performing an LC-MS/MS experiment on the combined sample. Relative protein abundances are then calculated from the abundances of the corresponding transitions.

There is however a danger that the selected peptides may not all be labeled. Partly for this reason the most common procedure for quantification using SRM is to use internal standards. This approach makes use of synthetic peptide standards, one for each peptide for which a transition will be monitored. These synthetic versions of the targeted peptides carry heavy stable isotopes, for example, ^{15}N and ^{13}C, to distinguish them by mass from the sample-derived peptides, and are spiked into the individual samples. Separate LC-MS/MS runs are then performed on each sample

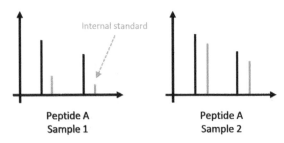

Peptide A   Peptide A
Sample 1   Sample 2

Figure 15.7   Label based quantification using SRM. The same peptide is found in two separate samples and compared to a common internal reference or standard (typically a synthetic version of the peptide carrying heavy isotopes), illustrated by the grey peaks. The amount of the internal standard is equal in both samples. In this illustration, it is therefore clear that the peptide is much more abundant in Sample 1 than in Sample 2, a conclusion that would not have been obvious from a direct comparison of the black transition peaks.

with spiked-in standards. The heavy stable isotope-carrying internal standards are often referred to as *heavy peptides.*

The abundance of an unlabeled sample peptide relative to its corresponding labeled standard peptide is calculated from the relative intensities of the corresponding transitions, and the relative peptide abundance between two samples is calculated from the two standard relative abundances. Figure 15.7 illustrates this.

The labeling has to introduce a sufficiently large mass difference such that the two peptide variants will not interfere. Most standard peptide labeling techniques can be used for this purpose, for example, ICAT and SILAC, provided the introduced mass shift is large enough.

Another important benefit of heavy peptides is that they can be used to validate whether the signals observed for a transition really come from the selected peptide/fragment. This is further discussed in Section 15.7.

Again we point out that to reduce the variability in the processing of the internal standard and its corresponding target peptide, the standards ought to be incorporated into the sample as early as possible. For this purpose it would therefore be best to spike in the standards before digestion, but as the standards are peptides and not proteins this is rarely done.

A labeling technique designed especially for SRM is *mTRAQ.* In mTRAQ the labels are an extension of the iTRAQ labeling technology, in that the same amine reactive group and a similar structure are used for the reagent. Note however, that the mTRAQ labels, unlike the iTRAQ labels, are nonisobaric. The peptides labeled with the different mTRAQ labels have identical retention time and ionization characteristics but different masses (0, 4, or 8 Da difference for triplex labeled Lys peptides and 0, 8, or 16 Da difference for Arg peptides). SRM quantification is performed using the fragment ions instead of the reporter ions for higher specificity in complex mixtures.

15.5   Requirements for SRM transitions

Reliable quantification using SRM depends strongly on the selection of the transitions for the targeted proteins, thus highlighting the importance of selecting the best possible transitions. This in turn results in several requirements for both the peptides as well as their fragment ions.

15.5.1   Requirements for the peptides

The total number of possible tryptic peptides for a protein ranges from tens to hundreds (53 on average for the human complement of UniProtKB/Swiss-Prot), but only a small subset of these peptides are routinely observed. It is therefore essential to select the most suitable peptides, satisfying the following requirements:

- The (peptide precursor) mass should have high specificity for the protein. Ideally the mass (or more precisely the m/z-value) should uniquely infer the protein in the considered context, that is, being a *proteotypic* or *signature* peptide.

- The peptide should have a high likelihood of being ionized.

- It should primarily ionize in one specific charge state.

- It should produce good MS/MS spectra upon fragmentation.

- It should not be easily modified, unless a modified form of a peptide is specifically targeted.

- It should be easily and completely cleaved at the tryptic cleavage sites, and not at other sites.

- Depending on the goal, care should be taken regarding isoform resolution: If specific (splice) isoforms are to be quantified, the peptide must distinguish this specific isoform; if, on the other hand, overall gene expression is to be quantified, the peptide should not distinguish between isoforms.

- The peptide should elute from the chromatographic column in a single, limited time interval.

There are groups of peptides that are known to not satisfy some of the above requirements and these peptides should thus be avoided:

- Peptides with flanking KK, KR, RR, RK residues (if using trypsin), due to the likelihood of missed cleavages.

- Peptides containing M, or W, because they readily oxidize.

- Peptides containing C, because they can be part of disulfide bonds.

- Very hydrophilic or hydrophobic peptides, since they elute too quickly, too late, or not at all from the LC column.

- Peptides that are too short or too long, that is, shorter than 6 or longer than 20 residues.

- Peptides with possible (internal) missed cleavages.

- Peptides with Q and N, due to possible deamidation.

- Peptides with N-terminal Q, as these are quickly transformed to pyro-glutamate under acidic conditions.

- Peptides containing a P. Since the peptide bond of P is generally quite weak, it is the bond that fragments most readily, leading to a large peak for the corresponding fragment ion, and strongly reduced peaks for all other fragment ions. Sometimes this behavior is used to advantage, since the peptide fragmentation is thus quite predictable, yet this is offset by the presence of only one reliable peak in the fragmentation spectrum, where typically multiple peaks are used to obtain reliable quantification.

To get reliable results, 3–5 peptides ought to be selected per protein.

15.5.2   Requirements for the fragment ions

Requirements for ensuring good fragment ions are:

- The fragment ion mass together with the peptide precursor mass should uniquely identify the considered protein in the experimental context.

- It should be one of the most intense peaks in the MS/MS spectrum for the peptide.

- It should have a low intensity variance across multiple MS/MS spectra of the same peptide with the same quantity.

- The observed intensity in the MS/MS spectra should depend linearly on the abundance of the protein in the sample.

- The limit of detection (LOD) and the limit of quantification (LOQ) should be low, thus making it possible to quantify low abundance proteins.

- The mass (m/z) of the fragment ion should be larger than the m/z of the precursor peptide, since larger fragment ions are typically more readily observed.

- The peak should not interfere with other peaks.

Two to five fragment ions per peptide are usually selected, and note that this means that the first requirement (that a single transition is unique for a peptide, and thus a protein) can be replaced by a more readily satisfied constraint: The set of selected transitions should be unique for a peptide, and thus for a protein. Also note that the linearity of the intensity response to the amount of peptide in the sample is dependent primarily on the linear performance range of the instrument detector, and may therefore not be determined by the fragment ion proper.

15.6   Finding optimal transitions

The process of finding transitions that satisfy the above listed requirements can be summarized as follows:

1. Select the proteins to be quantified.

2. For each selected protein, find the peptides that are most suitable for SRM.

3. For each selected peptide, find the fragment ions that are most suitable for SRM.

Figure 15.8 illustrates the process.

It is important to realize that the satisfaction of the requirements depends on the sample, the equipment used, and on experimental parameters. This means that transitions that worked well in one experiment may not work well in another experiment. However, knowing the good transitions from other experiments may still help the user in selecting transitions for the actual experiment, if only to establish certain patterns that can be useful in the selection process.

For this purpose, databases of transitions are compiled, with SRMAtlas as the best-known example. Tools have also been developed for predicting good transitions, and popular examples are given in the Bibliographic notes. However, in almost all cases transitions first have to be experimentally validated in the actual experimental setup before use.

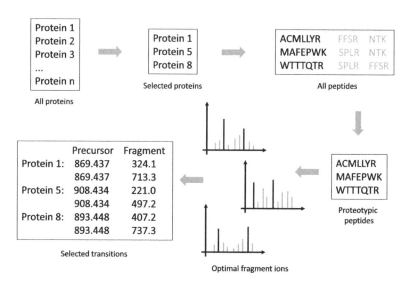

Figure 15.8   Selecting transitions: (i) a list of proteins to analyze is selected, (ii) the possible peptides for the selected proteins are listed, (iii) the protein's proteotypic peptides are found, (iv) the optimal fragment ions for each peptide are determined, and (v) the actual transition list is compiled.

15.7   Validating transitions

Validation here relates to the thorough analysis of SRM experiments (most often carried out using heavy labeled internal standards) to verify that the selected transitions can be used as expected. This process includes several steps that can be performed in different orders.

15.7.1   Testing linearity

The end result is to find the relative abundance of a peptide p in two samples using a shared internal standard S (a labeled variant of p).

Let

- $p_{r,1}$, $p_{r,2}$ be the real abundances of p in the two samples respectively;

- $p_{c,1}$, $p_{c,2}$ be the calculated abundances of p in the two samples respectively;

- S_r be the real abundance of S;

- $S_{c,1}$, $S_{c,2}$ be the calculated abundances of S in the two samples respectively.

Then the relative abundance R_p is calculated as

$$R_p = \frac{\frac{p_{c,1}}{S_{c,1}}}{\frac{p_{c,2}}{S_{c,2}}}.$$

R_p thus provides an estimate for $\frac{p_{r,1}}{p_{r,2}}$.

For this to work the requirements mentioned in Chapter 9 for labeled quantification must be satisfied. In addition $\frac{p_{c,i}}{S_{c,i}}$ must depend linearly on $p_{r,i}$, $i = 1, 2$.

Example Assume that we have $p_{r,1} = 100$, $p_{r,2} = 100$, and $\frac{p_{c,1}}{S_{c,1}} / \frac{p_{c,2}}{S_{c,2}} = 1$. If $p_{r,1}$ is changed to 300, $\frac{p_{c,1}}{S_{c,1}} / \frac{p_{c,2}}{S_{c,2}}$ should be 3. Since the only change in this second calculation is in $\frac{p_{c,1}}{S_{c,1}}$, this fraction should be three times larger than the corresponding fraction in the first calculation.
$\triangle$

Thus we should perform several experiments with varying abundances of p for a fixed abundance of S, and show that $\frac{p_{c,i}}{S_{c,i}}$ depends linearly on $p_{r,i}$. However this is equivalent to keeping the abundance of p fixed while varying the abundance of S. Since S is more easily spiked-in at various concentrations, varying the abundance of S is generally used. However, since the linear response range of the instrument's detector is also intrinsically limited, it is not simply a matter of verifying that $\frac{p_{c,i}}{S_{c,i}}$ depends linearly on $p_{r,i}$. Rather, the art consists of choosing the right S_r to spike into the sample, so that $\frac{p_{c,i}}{S_{c,i}}$ falls within the linear range of the detector. Note that this can be a serious challenge if the abundance of p varies a lot between samples, that is, when $\frac{p_{r,i}}{p_{r,j}}$ is very large or very small.

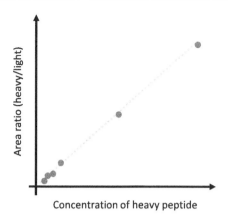

Figure 15.9   A response curve, showing the ratio of internal standard to peptide (heavy/light) for different abundances of the internal standard. Here, the ratio is more or less linear across the entire range.

The results from the different experiments are often presented in a plot with S_r along the horizontal axis, and $\frac{S_{c,1}}{p_{c,1}}$ (or more often the logarithm of this ratio) along the vertical axis. This curve is called a *response curve*. Regression analysis can then be used to test for linearity, and identify the interval for S_r where linearity is achieved. Note that because of the detector's linear range limitations, linearity is most readily achieved around a value of 1 for $\frac{S_{c,1}}{p_{c,1}}$.

Figure 15.9 shows an example for such a response curve. Response curves are also referred to as *loading plots*.

15.7.2   Determining retention time

For each of the peptides the retention time should either be predicted or be determined from other experiments, such that scheduled SRM can be employed.

15.7.3   Limit of detection/quantification

Limit of detection can be determined as described in Chapter 5.4. However, the signal-to-noise ratio can also be used as a proxy, for example by defining $LOD = 3\frac{S}{N}$, and setting $LOQ = 3LOD$.

15.7.4   Dealing with low abundant proteins

One way of dealing with low abundant proteins is to fractionate the peptide sample, such that low abundant peptides are not overshadowed by high abundant peptides. The optimal fractionation procedure is usually based on the experience gained from previous experiments.

15.7.5   Checking for interference

The transitions should be checked for interference from other transitions or noise. There are various ways in which this can be performed, with the following being the most common examples.

Compare the SICs from a peptide   The intensities for the transitions of the same peptide are acquired 'simultaneously.' So even though their intensities can vary, the transitions should all have the same peak shape in the SIC. If one transition deviates from the others in terms of its peak shape, it is most likely not derived from the targeted peptide, or suffers from noise interference.

Using MS/MS spectra for identification   The acquisition of a transition can be followed by a full fragment ion scan, effectively producing an MS/MS spectrum. This MS/MS spectrum can then be searched against a protein sequence database to verify the peptide identity. A drawback with this approach is a reduced dwell time for the transitions, thus decreasing sensitivity and accuracy.

Score the SICs from a peptide   In *mProphet* all the peaks in the SICs corresponding to a peptide are considered a 'unit,' and given a score depending on the 'likelihood' that they come from a real peptide. The scores of all targeted peptides are sorted, and a false discovery rate (FDR) is calculated based on a decoy transition list, in the same way as described in Chapter 4.7. A decoy transition is made for each transition.

Comparing SICs from the labeled/unlabeled peptide   In label based SRM quantification it is possible to validate transitions by employing the fact that the light and the heavy versions of the fragment ions from the same peptide ought to have the same relationships between the peak intensities. The actual intensities of the peaks will however vary due to minor variations in the experiment. For the intensities in the diagrams in Figure 15.10 we should ideally have $a/e = b/f = c/g = d/h$. We see that d/h clearly does not satisfy this condition. It is therefore most likely subject to contamination, and should be excluded from further analysis.

Manual validation   Various forms of validation can be employed by manually inspecting the transitions and spectra. Manual validation is however not recommended as this is a time consuming task, and it is also difficult to reproducibly select valid transitions unless very strict rules are employed.

15.8   Assay development

To develop SRM assays means to come up with a detailed procedure that can be followed to quantify a given set of proteins through selected transitions. This includes (but is not limited to) describing the optimal set of peptides and fragment ions to use for a given sample type and recommending an optimal instrument setup.

Most often the goal of SRM assay development is to validate potential biomarker candidates prior to further in-depth (and usually expensive) analysis. Setting up the

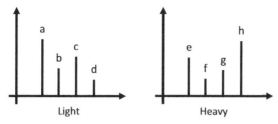

Figure 15.10   Two 'spectra' constructed from the SICs of four corresponding transitions from the light and heavy version of a peptide, illustrating contamination of d,h.

assays is however a long process in itself. The general workflow for developing such assays for biomarker discovery can be described as follows:

1. Perform discovery quantification to discover possible candidate proteins for biomarkers.

2. Perform targeted quantification to validate the candidates.

3. Perform pre-clinical and clinical validation.

For more on biomarkers see Chapter 18.

While there are exceptions, like the SRMAtlas, there are still very few publicly available SRM assays. Also the databases that do exist are still either very limited or restricted to specific organisms and/or experimental setups. Developing and validating SRM assays therefore remains both a labor and time consuming task. However, once assays have been established for a particular sample type and experimental setup, it is straightforward to repeat the method for multiple samples, allowing SRM to be considered a high-throughput method.

As mentioned before there are various software tools that can be used for SRM assay development. Some tools allow you to define the whole pipeline, while others focus on one or more elements of the pipeline. A review of the most common tools can be found in the Bibliographic notes.

A lot of work is currently carried out in simplifying the exchange of SRM data between the various SRM tools and instruments. An important step in this direction is the development of a standard data format for SRM transitions, called TraML. See Chapter 19 for details.

15.9   Exercises

1. Consider an imagined SRM-run of 20 transitions, and a dwell time of 20 ms. For a given protein we consider two peptides called peptide 1 and peptide 2, for which we select two transitions each. We define the time when the cycling starts as time 0. Peptide 1 elutes during the time interval from 2 min 20 sec to 3 min 10 sec, and

peptide 2 from 4 min 8 sec to 4 min 40 sec. How long is the total time used to detect signals for these peptides?

2. We have performed SRM on two samples, and consider a protein for which we have two peptides with two transitions each. The observed abundances are:

	Sample 1	Sample 2
Peptide 1, transition 1	802	1464
Peptide 1, transition 2	404	738
Peptide 2, transition 1	123	218
Peptide 2, transition 2	758	1360
Sample Total	70 620	60 484

Calculate the relative abundance for the protein between the two samples.

15.10   Bibliographic notes

Limit of detection/quantification   Linnet and Kondratovich (2004), Armbruster and Pry (2008), Keshishian et al. (2007), Keshishian et al. (2009).

SRM tutorial   Lange et al. (2008).

mProphet   Reiter et al. (2011).

mTRAQ   DeSouza et al. (2008), www.sciex.com/products/standards-and-reagents/mtraq-reagents.

SRMAtlas   www.srmatlas.org.

Tools   Review: Cham et al. (2010).

HUPO-HPP   www.hupo.org/research/hpp.

TraML   Deutsch et al. (2012), www.psidev.info.

Instrument details   www.thermoscientific.com and www.absciex.com.

16

Absolute quantification

Although relative quantification is a very useful tool for the comparison of different samples, it is often necessary to establish the absolute quantity of a peptide or protein in a sample. An example of this is the search for biomarkers, where the absolute abundance of a peptide biomarker can provide useful information about the suitability of different assays for detecting this peptide in subsequent diagnostic procedures. Absolute quantification is also needed for the modeling of biological systems and for understanding the complex biochemical processes in a cell.

By absolute quantification we here mean the determination of the concentration of a protein in a sample, or, by derivation, the amount of protein in a cell. The absolute abundance of a protein is usually given as mole or mole/volume.

16.1 Performing absolute quantification

The general procedure for absolute quantification is to use internal standards for which the absolute abundances are known. These standards can be peptides, a concatenation of peptides, or whole proteins, depending on the method used. However, the last method described in this chapter does not use standards at all.

When standards are employed, the overall strategy is always quite similar: The absolute abundance of a peptide or protein is determined by the relative abundance between this peptide or protein and the corresponding internal standard. Most of the techniques for relative quantification described earlier in this book can therefore be used to obtain absolute quantification information in combination with an internal standard.

It is clear that this strategy depends completely on accurately knowing the quantity of the standard added to the sample, and on the purity of that standard. The quantity of the standard is typically determined by the producer. Do note that subsequent handling

Computational and Statistical Methods for Protein Quantification by Mass Spectrometry,
First Edition. Ingvar Eidhammer, Harald Barsnes, Geir Egil Eide and Lennart Martens.
© 2013 John Wiley & Sons, Ltd. Published 2013 by John Wiley & Sons, Ltd.

steps, including redissolving the internal standard, pipetting, or freeze–thaw cycling, may alter the actual quantity of the internal standard peptide, leading to subsequent errors. Handling should therefore be minimized as much as possible to maintain the accuracy of the abundance estimate of the producer.

16.1.1 Linear dependency between the calculated and the real abundances

In order to calculate the real absolute abundance of a protein from the calculated abundances there must be a linear relationship between the calculated and the real abundances.

Let:

- S_r and S_c be the real and the calculated abundances of the standard respectively;

- P_r and P_c be the real and the calculated abundances of the peptide/protein under consideration.

Then P_r should be calculated from the known S_r, and from the calculated S_c and P_c. When $P_r = S_r$ the calculated values should be equal, such that

$$\frac{P_r}{S_r} = \frac{P_c}{S_c} = 1.$$

When P_r is increasing or decreasing, these two ratios should also increase or decrease in the same ratio (on the assumption that S_r remains fixed), such that the formula for determining P_r becomes:

$$P_r = S_r \frac{P_c}{S_c}. \tag{16.1}$$

Thus one must show that this formula holds for the standard and the considered peptide or protein. This is typically verified by performing different experiments with fixed standard abundance and varying abundances for the peptide or protein, plotting the results in a scatter plot, and calculating the regression line through the results (in chemistry this is called a *titration curve*).

Given that the dynamic range can be large, the ratios are most often represented as log ratios, as shown in Figure 16.1. Note that for the regression line $y = ax + b$ to give correct calculation of the abundance, b should be zero and a not deviate too much from one.

16.2 Label based absolute quantification

Label based approaches are more accurate than label free approaches, and are therefore also more common. The label based approaches are considered to be targeted, in the sense that one must know in advance which proteins to create standards for. A

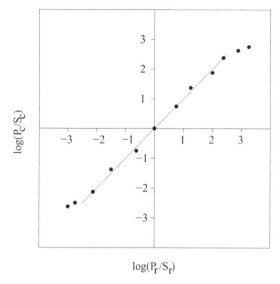

Figure 16.1   Illustration of the requirement for linearity between the ratios. Assuming base 10 for the logarithm, the figure shows linearity in a dynamic range of over four orders of magnitude.

standard for a protein or peptide is a labeled version of the protein or peptide, ensuring that the target and the standard have the same chemical and physical properties in the experiments, except for their mass.

As always, the standards should be incorporated into the samples as early as possible to reduce variability in the handling of the standard and the target. As usual, using multiple peptide standards for a protein will increase the accuracy as it increases the number of measurements.

The label based methods can be divided into three groups, depending on the type of standard used:

- Stable isotope-labeled peptide standards.

- Stable isotope-labeled concatenated peptide standards.

- Stable isotope-labeled intact protein standards.

There are also various methods relying on a combination of the above.

16.2.1   Stable isotope-labeled peptide standards

These methods are usually employed in quantification assays by SRM, but can be used for more traditional MS analyses as well. One or several representative peptides are selected for each of the proteins to be quantified, and synthetic labeled peptides carrying heavy isotopes are produced. A fixed abundance of these labeled peptides are then mixed into the sample to be quantified (preferably into the protein sample), and SRM is performed. A well known protocol employing this strategy is AQUA (Absolute QUAntification).

16.2.2   Stable isotope-labeled concatenated peptide standards

Since the purity of synthetic peptides is variable, and the handling can vary, unwanted variability and errors can be introduced in the analysis. To remedy this, methods for synthesizing a polypeptide consisting of the peptides of interest from across all the proteins to be quantified have been developed.

One such technique is called QCAT (or QconCAT), where the concatenated polypeptide is called a QCAT protein. One representative peptide sequence from each protein is selected and concatenated with the other selected peptides. DNA is then synthesized representing the polypeptide chain obtained after concatenation. When this nucleotide sequence is added into an expression vector and thereafter transfected into a bacterium, the resulting protein will yield the desired peptides upon enzymatic (usually tryptic) cleavage. By growing these bacteria on a medium that contains one or more heavy amino acids, the QCAT proteins (and thus their peptides) will be labeled. The process of generating the QCAT protein is illustrated in Figure 16.2.

A fixed (and known) amount of the QCAT protein is mixed with the sample containing the proteins to be quantified, and digestion and LC-MS/MS analysis are subsequently performed. The output for each peptide in this analysis is a ratio between the target peptide intensity and the QCAT peptide intensity. Given that the absolute abundance of the QCAT protein is known, the absolute abundance of the target peptide can be derived, in turn allowing extrapolation of the protein abundance. A drawback with this method is that the protein abundance is calculated from a single peptide. To increase the reliability of the quantification two (or more) peptides can be selected for each protein.

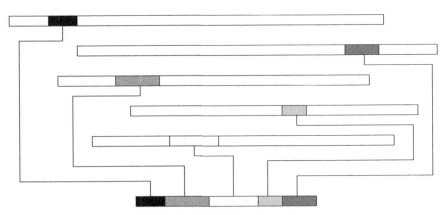

Figure 16.2   Schematic overview of the generation of a synthetic polypeptide designed to quantify five different proteins. One peptide from each protein is selected, and these are concatenated into the QCAT protein.

An alternative to QCAT is PCS (Peptide Concatenated Standard). In contrast to QCAT the peptides here also include some of the flanking amino acids surrounding the cleavage site of the enzyme. In this way the cleavage efficiency of the original proteins will be maintained, and the accuracy of the quantification is improved because any potential cleavage losses (e.g., due to missed cleavages) in the original proteins are more accurately mimicked by the concatenated peptides.

16.2.3   Stable isotope-labeled intact protein standards

To remedy the shortcomings of using peptides one can instead use intact labeled proteins as standards. Regular relative label based quantification can be performed, but with a specific focus on the peptides that originated from the proteins under consideration (for which masses and sequences are known).

One protocol that relies on this approach is Absolute SILAC, using MaxQuant for the relative quantification. A recently developed variant of Absolute SILAC makes use of PrESTs (Protein Epitope Signature Tags) as standards. A PrEST is a subsequence of a target protein of about 100 amino acids with minimal homology to other proteins. Signal peptide sequences are also avoided, as are membrane spanning regions.[1]

Alternatives to using SILAC are PSAQ (Protein Standard Absolute Quantification), and FlexiQuant.

16.3   Label free absolute quantification

The label free methods use proteins as standards. Note however that the standards are different proteins from those to be quantified. These methods can be grouped depending on which spectra are used for quantification: MS spectra or MS/MS spectra.

16.3.1   Quantification by MS spectra

Peak intensities are used for quantification, and we describe two methods.

Using the three most intense peptides – top3

A simple method for the determination of absolute abundance relies on utilizing a feature of the relationship between the number of molecules in the sample and the calculated abundances of the three most abundant peptides in an LC-MS/MS run. This procedure is also referred to as *top3*.

The average abundance (signal response) of the three most intense tryptic peptides per mole of protein has been shown to be constant within a coefficient of variation of ±10%, Silva et al. (2006). When this is known, the protein abundances in a sample

---

[1] The PrESTs are produced by the Human Protein Atlas project as potential epitopes for the antibodies they produce.

can be calculated by using Equation 16.1, where P_c, S_c are computed as the average of the three most intense peptides in the protein and the standard, respectively.

Finding the three most abundant peptides   The first challenge in top3 is to determine and quantify the three most abundant peptides of the proteins to quantify. In ordinary LC-MS/MS the peptides to be fragmented are either determined by a list of predefined mass/retention time pairs (inclusion list), or the k (typically 3–5) most intense peaks in the MS spectrum are selected for MS/MS analysis (data dependent acquisition). This means that it is not guaranteed that the three most abundant peptides are found for each of the actual proteins, or even for the standards. Another type of acquisition is therefore used, termed *LC/MS^E acquisition.*

With LC/MS^E acquisition, all co-eluting peptides are fragmented together. This is done by a series of alternating low energy MS scans and high energy MS^E scans (E for elevated) along the total elution profile. Each scan typically runs for 2 seconds. In the MS scans MS spectra are created, and in the MS^E scans fragmentation occurs and MS/MS spectra are produced. Note however that each MS/MS spectrum contains a superposition of *all MS/MS spectra from all peptides eluting at the given retention time*. The next challenge is therefore to split these superimposed MS/MS spectra into separate MS/MS spectra for each peptide. This deconvolution of the combined spectra into individual MS/MS spectra is necessary for peptide identification.

The splitting of the compound MS/MS spectrum is performed by constructing an XIC for each small mass interval, both at the MS level and at the MS/MS level. All fragments in an MS/MS spectrum belonging to a single precursor should have similar XIC-profiles, (in the same retention interval) as that precursor. Thus by comparing the XICs, the fragments belonging to each peptide can be determined, as illustrated in Figure 16.3, and a peptide specific MS/MS spectrum can be constructed to identify the precursor peptide. Note however that interference between the fragments from different peptides typically renders this splitting rather complex, making it a nontrivial task.

16.3.1.1 iBAQ

iBAQ (intensity-Based Absolute Quantification of proteins) is a method where the intensity of a protein is calculated as the sum of all identified peptide intensities. This intensity is then divided by the number of theoretically observable peptides, which are considered to be all peptides with a length from 6 to 30 residues. The resulting intensities are called iBAQ intensities, and represent the calculated protein abundances. Several standards are used, and the logarithmically transformed real and calculated abundances of the standards are plotted in a scatter plot, with the calculated value on the horizontal axis. A regression line is found, and used to calculate the real abundances of all identified proteins using their iBAQ intensities. An advantage of this method is that the proteins to be quantified do not have to be known *a priori.*

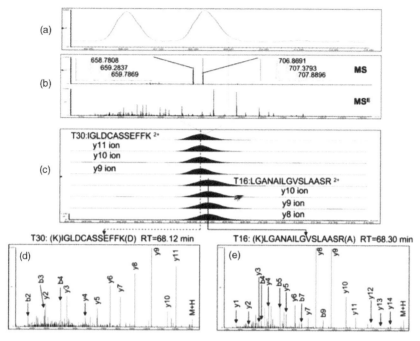

Figure 16.3   Illustration of the process of splitting an MS/MS (MS^E) spectrum. A is part of a chromatogram showing a peak around 68.3 min. B shows examples of MS and MS/MS spectra from this time interval. The MS spectrum reveals peaks from two peptides with masses 659 and 707. C shows eight XICs, two from the MS spectra and six from the MS/MS spectra. D and E are the deconvoluted MS/MS spectra for the two peptides. From the XICs it is easy to see which peaks (fragments) belong to which peptide. Reproduced from Chakraborty et al. (2007) by permission of John Wiley & Sons.

16.3.2   Quantification by the number of MS/MS spectra

This strategy relies on quantification by peptide identification, described in Chapter 14, and is the only method described here that does not rely on standards. However, when linearity (or other known dependencies) exists between the calculated values of the indices and the absolute abundances, standards of known absolute abundances can be employed as usual.

APEX (Absolute Protein EXpression measurement), described in Lu et al. (2007), calculates an absolute abundance for a protein based on:

$$APEX_i = \frac{n_i\, p_i}{o_i \sum_{k=1}^{I} \frac{n_k\, p_k}{o_k}} C,$$

where

- n_i is the total number of MS/MS spectra observed for peptides of protein i;
- p_i is the probability that the protein is correctly inferred, as returned by the database search program;
- o_i is the expected number of unique peptides that should be observed for protein i;
- I is the number of inferred proteins in the sample;
- C is the total number of protein molecules in the sample.

n_i, p_i and I are calculated by the software, C is estimated, and o_i is a function of the ionization efficiency of the peptides, the solvent conditions, the m/z of the peptides, and some additional factors. A classifier for predicting observable peptides is therefore developed from test data in order to obtain o_i.

$APEX_i$ thus directly calculates an estimate for the absolute abundance of protein i in the sample.

16.4   Exercises

1. We consider label free quantification by MS spectra. XICs are constructed during a given time interval, determining the intensities by 12 measurements. Assume that the following XICs are produced for two peptides and four fragments:

P_1	423	675	2430	4056	4123	2768	940	260	312	290	335	256
P_2	234	312	321	892	2002	4326	5238	5098	2679	967	291	287
F_1	312	584	1834	2856	2798	1846	720	256	245	287	302	287
F_2	252	299	301	804	1867	3980	4667	4623	2404	801	301	256
F_3	278	301	323	634	1602	3456	4234	4302	2019	713	297	263
F_4	324	367	2156	3402	3489	2434	867	278	289	267	278	234

Determine to which peptide each fragment belongs.

16.5   Bibliographic notes

Label based peptide standards   Gerber et al. (2003)(AQUA), Kirkpatrick et al. (2005). Kettenbach et al. (2011).

Label based concatenated peptide standards   Beynon et al. (2005)(QCAT), Kito et al. (2007)(PCS).

Label based protein standards   Hanke et al. (2008); Zeiler et al. (2011)(Absolute SILAC), Geiger et al. (2010, 2011)(SILAC), Brun et al. (2007)(PSAQ), Singh et al. (2009)(FLEXIQuant).

Label based combined Ishiama et al. (2005b)(CDIT), http://www
.proteomefactory.com (MeCAT).

Label free MS based Silva et al. (2006), Chakraborty et al. (2007), Cheng et al.
(2009), Schwanhäusser et al. (2011)(iBAQ).

Label free MS/MS based Lu et al. (2007).

Combined methods Malmström et al. (2009).

17

Quantification of post-translational modifications

Post-translational modifications (PTMs) are crucial for the heterogeneity of proteins. Indeed, since PTMs can change a protein's structure and function, a given function of a protein can be switched on or off, or a single protein can be used for different cellular functions in different tissues. Post-translational modifications are therefore often used to regulate how proteins will act in eukaryotic organisms, and disregulation of PTMs is important in many diseases. Examining how the extent of modifications varies over different situations is therefore interesting from both a basic biological as well as from a clinical perspective.

17.1   PTM and mass spectrometry

Although mass spectrometry has been applied to the quantification of PTMs for more than ten years, there are essentially no general methods available that can analyze all types of modifications. Rather, methods have been developed for the analysis of specific types of modifications, primarily because most modifications are sufficiently uncommon and require specific enrichment prior to mass spectrometry analysis.

Perhaps the most studied PTM in proteomics is phosphorylation, where many different enrichment and analysis protocols have been developed, and are routinely applied. Other modifications pose more significant problems, including ubiquitination, lipidation, and glycosylation. The latter are especially difficult to analyze, since glycosylation involves the addition of branched trees of several glycans that

Computational and Statistical Methods for Protein Quantification by Mass Spectrometry,
First Edition. Ingvar Eidhammer, Harald Barsnes, Geir Egil Eide and Lennart Martens.
© 2013 John Wiley & Sons, Ltd. Published 2013 by John Wiley & Sons, Ltd.

are unstable in the mass spectrometer, creating highly complex and essentially unpredictable patterns in the corresponding fragmentation spectra.

For those modifications (such as phosphorylation) that can already be analyzed using high throughput mass spectrometry, there are additional challenges, most notably the precise localization of the modification on the peptide, and thus the protein. The problem is that many modifications are unstable during fragmentation, resulting in the loss of the modification from (some of) the fragment ions. Some approaches to analyze modifications require the enzymatic or chemical removal of the modification during enrichment, effectively rendering site localization impossible (at least, if multiple potential modification sites are found in the peptide).

17.2   Modification degree

Each type of modification can occur at specific amino acid residues, and/or at specific positions, for example, acetylation at the N-terminus of a protein. When considering a protein P and a (type of) modification μ we say that each residue at which μ can occur is a *modification site* for μ in P.

For a specific sample (or situation) let:

- s be a modification site for μ in P;
- $P_\circ$ be the abundance of P in the sample;
- P_μ be the abundance of P where s is modified by μ.

Then define the *modification degree* (of s) for μ as

$$\frac{P_\mu}{P_\circ} \text{ or } \frac{P_\mu}{P_\circ} \cdot 100\%.$$

The modification degree can also be seen in terms of the *stoichiometry* in the following reaction:

$$P + \mu \rightleftharpoons P_\mu.$$

Note that each modification site of a protein can have a different modification degree for μ. The extent of the modification can depend on:

- the local structure around the site;
- the position of the site;
- the flanking residues of the site;
- the state of other modifiable residues.

We may want to determine two types of modification degrees:

- absolute modification degree is calculated for one sample or situation;
- relative modification degree is calculated for two samples or situations.

17.3   Absolute modification degree

Absolute modification degrees have to be calculated from the *observed* abundances. An observed abundance I_o is related to the real abundance P_o through $I_o = \rho_o P_o$, where ρ_o is called the *observability*. The observability depends on the degree of ionization of the peptide, and the detectability of the resulting peptide ions in the mass spectrometer.

An obvious way to determine the absolute modification degree of a site s of a protein is to consider a peptide q containing s. We then consider the peptide copies modified by μ at s, and those not modified, as two versions of the same peptide. The previously explained methods can then be used to search for variants and subsequent abundance determination.

We now consider a peptide q containing s, and let:

- q_μ be the abundance of the modified occurrences of q in the considered sample;

- q_n be the abundance of the nonmodified occurrences of q;

- $q_o = q_\mu + q_n$;

- I_μ, I_n, I_o be the observed (calculated) abundances respectively;

- ρ_μ, ρ_n, ρ_o be the observability values.

The modification degree md is then calculated by Equation 17.1.

$$md = \frac{q_\mu}{q_o} = \frac{\frac{I_\mu}{\rho_\mu}}{\frac{I_o}{\rho_o}} = \frac{\frac{I_\mu}{\rho_\mu}}{\frac{I_\mu}{\rho_\mu} + \frac{I_n}{\rho_n}}. \tag{17.1}$$

The abundances of the modified and unmodified versions of q must be determined separately, so the expression to the right must be used. However, the observability values are unknown. If we could assume $\rho_\mu = \rho_n$, md could be calculated as

$$md = \frac{I_\mu}{I_\mu + I_n} \left(= \frac{I_o - I_n}{I_o} \right).$$

But such an equality cannot be assumed. Thus other (often modification specific) methods are used. We here briefly describe three different principles. All consider a peptide containing s, and in the description we assume that μ at s is the only modification of the peptide. However, not all methods have this restriction.

17.3.1   Reversing the modification

If the modification is reversible, removing the modification can be used to calculate the modification degree. The idea is to split the sample into two aliquots, A_1 and A_2. In A_1 the modifications are removed, transforming all modified peptides back into unmodified peptides. Since md can be calculated as the ratio of the number of modified forms of a peptide over the total count of this peptide, we can use the observed abundances in the aliquots to derive md as follows. I_o corresponds to the sum of the observed abundance of the originally unmodified and modified peptides, and is now represented by the abundance of the unmodified peptide in A_1. I_n on the other hand, is given by the abundance of the unmodified peptide in A_2.

Since we only consider the unmodified peptide in both aliquots, the observability values will now all be equal. The modification degree is then calculated from these two abundances by Equation 17.2.

$$\frac{I_o - I_n}{I_o}. \tag{17.2}$$

Below we sketch two methods for determining phosphorylation degrees where a phosphatase is used for removing the phosphate moiety.

A method using labeling

Hegeman et al. (2004) describe a method where the phosphatase treated aliquot is labeled by a heavy label, and the non-treated aliquot by a light label. These aliquots are then recombined after labeling, and LC-MS/MS is performed. Figure 17.1 illustrates the principle.

In this type of experiment, three peaks will occur in an MS spectrum of a phosphorylated peptide. The intensity I_n of peak a comes from the nonphosphorylated peptide occurrences of the light aliquot, and the intensity I_μ of peak c comes from the phosphorylated peptide occurrences of the light aliquot. The intensity I_o ($=I_n + I_\mu$) of peak b comes from the heavy labeled, phosphatase treated aliquot. Note that the peak at d, where phosphorylated heavy peptides should have occurred, is empty because of the removal of all phosphate moieties by the phosphatase. The degree of phosphorylation can then be calculated using Equation 17.2.

A method using selected reaction monitoring

Domanski et al. (2010) describe a method to determine the degree of phosphorylation for a given protein using SRM. Here the sample is split into two aliquots, and in one of them the phosporylations are reversed. Heavy labeled standard peptides are added to both aliquots for each targeted peptide.

An SRM experiment is subsequently performed for both aliquots. Then

- aliquot 1 is used to calculate I_o and I_{S_1}, the observed abundance of the standard (heavy peptide);

- aliquot 2 is used to calculate I_n and I_{S_2}, the observed abundance of the standard (heavy peptide).

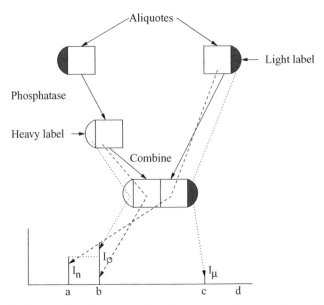

Figure 17.1   Illustration of the use of a phosphatase for calculating the degree of phosphorylation. The marked regions represent phosphorylated occurrences. The distance between peaks a and c is equal to the mass of the phosphate (when one charge is assumed). The distance between a and b is equal to the difference between the heavy and light label.

Then the modification degree of q is calculated by Equation 17.3.

$$md = \frac{\frac{I_\circ}{I_{S_1}} - \frac{I_n}{I_{S_2}}}{\frac{I_\circ}{I_{S_1}}}, \qquad (17.3)$$

where the standards are used to normalize the abundances between the two runs.

17.3.2   Use of two standards

In this method there is one standard for the modified version of the peptide, and one for the unmodified version. We extend the notation by:

- let I_μ^S, I_n^S be the observed abundances of modified and unmodified occurrences of the standards respectively.

We assume that the observability of the two versions of the standard are equal to the observability of the two versions of the peptide, respectively. By letting the abundances of those two standards be equal (to K), we can formulate

$$K = \frac{I_\mu^S}{\rho_\mu} = \frac{I_n^S}{\rho_n} \Rightarrow \frac{\rho_\mu}{\rho_n} = \frac{I_\mu^S}{I_n^S}.$$

Then the modification degree is calculated using Equation 17.1.

$$\frac{\frac{I_\mu}{\rho_\mu}}{\frac{I_\mu}{\rho_\mu} + \frac{I_n}{\rho_n}} = \frac{I_\mu}{I_\mu + \frac{\rho_\mu}{\rho_n} I_n} = \frac{I_\mu}{I_\mu + \frac{I_\mu^S}{I_n^S} I_n}.$$

17.3.3   Label free modification degree analysis

Steen et al. (2005) describe a method used for phosphorylation stoichiometry, but the principle can also be used for other types of modifications. To calculate absolute quantification two samples must be analyzed. Separate LC-MS/MS experiments (runs) are performed for each sample, since the method is label free. Note that normalization has to be performed among the different experiments/runs to make them comparable.

Let A, B be the two samples, that are used as superscript indices in the notations. The method relies on some assumptions:

- the increase in the number of modified occurrences of q results in an equal decrease of the number of unmodified occurrences, and vice versa;

- the observability values of the peptides are equal for the considered samples;

- the observed abundances are normalized such that $q_\circ$ of the two samples can be set equal;

- the difference in modification stoichiometry of q in the two situations must differ by at least 10%.

Letting I represent the normalized values we can then formulate

$$q_\mu^A = \frac{I_\mu^A}{\rho_\mu}, \; q_n^A = \frac{I_n^A}{\rho_n},$$

and similarly for B.
Further

$$\frac{I_\mu^A}{\rho_\mu} + \frac{I_n^A}{\rho_n} = \frac{I_\mu^B}{\rho_\mu} + \frac{I_n^B}{\rho_n}.$$

From this we obtain

$$\frac{\rho_\mu}{\rho_n} = \frac{I_\mu^B - I_\mu^A}{I_n^A - I_n^B}.$$

Thus the known observed abundances can be used to determine $\frac{\rho_\mu}{\rho_n}$, and an absolute value for the modification degree can be calculated by Equation 17.4.

$$\frac{q_\mu^A}{q_\mu^A + q_n^A} = \frac{\frac{I_\mu^A}{\rho_\mu}}{\frac{I_\mu^A}{\rho_\mu} + \frac{I_n^A}{\rho_n}} = \frac{I_\mu^A}{I_\mu^A + \frac{\rho_\mu}{\rho_n} I_n^A}. \tag{17.4}$$

The calculations can be extended to hold for several samples, taking into account that the absolute quantification of n samples requires $n + 1$ samples.

Some of the assumptions for this method are tested and verified by Balasubramaniam et al. (2010) for SRM analysis.

17.4   Relative modification degree

Relative modification degree can be calculated from absolute modification degrees, for example, from Equation 17.4, but often it is more appropriate to calculate the relative modification degree directly.

We now consider:

- a protein P with abundances $P_\circ^A$, $P_\circ^B$ in two considered situations/samples respectively;

- a modification site s in P for modification μ;

- P_μ^A, P_μ^B are the amounts of protein P that are modified by μ in site s in the two samples.

Then the modification degrees are $\frac{P_\mu^A}{P_\circ^A}$, $\frac{P_\mu^B}{P_\circ^B}$, and the relative modification degree can be calculated by Equation 17.5.

$$\frac{\frac{P_\mu^A}{P_\circ^A}}{\frac{P_\mu^B}{P_\circ^B}} = \frac{P_\mu^A P_\circ^B}{P_\circ^A P_\mu^B} = \frac{\frac{P_\mu^A}{P_\mu^B}}{\frac{P_\circ^A}{P_\circ^B}}. \tag{17.5}$$

We call the ratio $\frac{P_\mu^A}{P_\mu^B}$ the *modification ratio*, and $\frac{P_\circ^A}{P_\circ^B}$ the *overall ratio*.

Example Let $P_\circ^A = 100$, $P_\circ^B = 200$, $P_\mu^A = 50$, $P_\mu^B = 40$. Then half of the protein occurrences in situation A are modified, but only one fifth in situation B. $\frac{P_\mu^A / P_\mu^B}{P_\circ^A / P_\circ^B} = \frac{1.25}{0.5} = 2.5$.

$\triangle$

The two ratios can be determined by mass spectrometry, and the exact procedure depends on the approach used to measure the abundances.

17.5   Discovery based modification stoichiometry

The methods described above are all meant for the targeted quantification of modification degrees. There are however also methods for discovery based modification degree quantification. Discovery here means discovering proteins at which modification μ has happened, and for each site at which μ has happened to calculate the relative modification degree using Equation 17.5. One should then

- identify and quantify peptides where modification μ has happened, to calculate the modification ratios;

- identify and quantify peptides to use for calculating the overall ratios.

Thus to be able to discover and calculate the modification degree for a modification μ at site s of protein P we need:

- an identified peptide covering s, and this for both situations;

- identified peptides from P without modifications, also for both situations.

In principle all of the previously described discovery based methods can be used, but their actual applicability depends on the chemical/biological/experimental techniques employed to obtain the abundances and perform the calculation of the two ratios, and whether enough data can be achieved to perform sound statistical analysis. Note that the identification of modified as well as unmodified peptides requires the modification to be specified as a variable modification in the identification software.

Calculating the two ratios can be performed by separate LC-MS/MS experiments, or in the same experiment. In the following we give examples of these two approaches, both label based.

17.5.1   Separate LC-MS/MS experiments for modified and unmodified peptides

Zhang et al. (2003) describe a protocol for identification and quantification of N-glycoproteins. All glycoproteins are isolated and digested. Then all glycopeptides are isolated, and the glycopeptide samples from the two situations are labeled by heavy and light labels, respectively, before combination. A standard LC-MS/MS experiment is then performed, and the modification ratios for the identified glycopeptides calculated. It should be mentioned that the protocol is not straightforward, and very careful and accurate execution is paramount.

The original protein samples (or only the isolated glycoproteins) are then analyzed by LC-MS/MS in the same way to calculate the overall ratios for the glycoproteins. This protocol is used with different labeling techniques, for example, Zhou et al. (2009) have used it with iTRAQ.

17.5.2 Common LC-MS/MS experiment for modified and unmodified peptides

Liu et al. (2010) describe a method for quantification of N-glycoproteins using ^{18}O labeling where the mass shift between the peptide variants in the two situations is different for glycopeptides and nonglycopeptides. In this way quantification of glycopeptides and nonglycopeptides can be performed in the same LC-MS/MS runs by using the different mass shifts to recognize the variants of glycopeptides and nonglycopeptides. Standard methods can then be used to determine the modification ratios and the overall ratios.

It should however be mentioned that the use of two mass shifts make the spectra more complicated and thus more difficult to analyze. In the method mentioned above, the mass shifts are 4 Da and 6 Da, resulting in overlaps of isotope envelopes for peptides with two or more charges that have to be corrected for.

17.5.3 Reliable results and significant differences

To get reliable modification degrees sufficient data has to be available. For an overall ratio several peptides may be used per LC-MS/MS experiment, but for a modification ratio only one peptide is usually available. Technical and/or biological replicates should therefore be used to increase the statistical power.

Another issue is the determination of the significance of observed changes in modification degree between two situations. While this issue resembles the overall problem of finding significantly regulated proteins, it is more subtle and therefore harder to address. The main problem lies in establishing a baseline distribution of 'unchanging' degrees of modifications. For overall quantification, the usual approaches assume that the majority of peptides are not differentially expressed, providing various means to extract this core distribution from the overall data, in turn allowing significance boundaries to be set for an unchanging distribution that directly support tests against the null hypothesis that peptide ratios are unchanged between samples. In the case of modification degrees, the same assumption could be made, but here it is much less likely to be correct, since most modifications are likely to be variable and in flux.

An easy way to circumvent the problem is to not assign significance at all, but just to rank the changes in modification degree from high to low. Such a list can then be analyzed top to bottom for promising leads. Alternatively, a baseline cutoff for significance can be calculated based on the (assumed known) performance of the quantitative mass spectrometry method used. Any significant difference needs to at least be larger than the combined, propagated error intrinsic to the technique. Note that this latter approach requires knowledge about this intrinsic error, in turn assuming that this error is relatively consistent and has been found through analysis of previous, standard quantitative experiments. Finally, one could also use an across-replicate approach, where significance is tested through the repeated occurrence of a particular ratio over many replicates.

17.6   Exercises

1. Show how Equation 17.3 is derived.

2. Assume that the following normalized abundances are obtained for a protein using a label free modification degree analysis:

$$I_\mu^A = 2490, \ I_n^A = 2406, \ I_\mu^B = 1446, \ I_n^B = 4032.$$

Calculate the modification degree of the protein in sample A by using Equation 17.4.

17.7   Bibliographic notes

Absolute modification degree   Hegeman et al. (2004), Domanski et al. (2010), Steen et al. (2005), Jin et al. (2010), Balasubramaniam et al. (2010), Eissler et al. (2011).

Targeted   Beck et al. (2006).

Discovery label based   Zhang et al. (2003), Zhou et al. (2009), Liu et al. (2010).

Overview   Zee et al. (2011).

18

Biomarkers

A biomarker is typically considered as a particular biological molecule (or set of molecules) that indicates a particular biological or pathological state, or that can be used to assess the response of an organism to a treatment.

As discussed in previous chapters, quantitative proteomics experiments are often performed as a means to discover novel biomarkers, or to validate such potential, newly discovered biomarkers. The former is often undertaken using discovery oriented proteomics approaches, while the latter typically benefits from more targeted approaches.

18.1 Evaluation of potential biomarkers

Any potential biomarker should be tested for its suitability, as explained in Chapter 8.4, where the sensitivity and specificity of such tests are calculated as

$$Se = \frac{TP}{TP + FN}, \quad Sp = \frac{TN}{FP + TN}.$$

In this context the sensitivity is the chance of finding the disease when it is present, the specificity is the chance of correctly concluding the disease absent.

Two other highly useful concepts when considering potential biomarkers are *positive and negative predictive values* of a test. The positive predictive value (PPV) in this context indicates how often the disease (or condition) is present given a positive test result, while the negative predictive value (NPV) indicates how often the disease is absent given a negative test. They can be calculated as

$$PPV = \frac{TP}{TP + FP}, \quad NPV = \frac{TN}{TN + FN}.$$

Computational and Statistical Methods for Protein Quantification by Mass Spectrometry,
First Edition. Ingvar Eidhammer, Harald Barsnes, Geir Egil Eide and Lennart Martens.
© 2013 John Wiley & Sons, Ltd. Published 2013 by John Wiley & Sons, Ltd.

18.1.1   Taking disease prevalence into account

In the evaluation of biomarkers it is important to take the occurrence rate of the disease, d, referred to as the *disease prevalence*, into account. In general, diseases with very low prevalence are much harder to detect reliably than diseases with a high prevalence.

When including the prevalence, the PPV and NPV can be calculated using conditional probabilities and Bayes' theorem. Let

- $+$ be a positive test result for an individual;
- D signify that the individual effectively has the disease.

Then

- $P(D)$ is the probability that a random individual in the population effectively has the disease, which is equivalent to the prevalence d;
- $P(D|+)$ is the PPV;
- $P(+|D)$ is the sensitivity of the test;
- $P(+)$ is the probability of a random individual getting a positive test.

Using Bayes' theorem we get

$$PPV = P(D|+) = \frac{P(+|D)P(D)}{P(+)}.$$

$P(+)$ must be calculated by adding together those testing positive that have the disease, and those testing positive that do not have the disease. Thus

$$P(+) = P(+|D)P(D) + P(+|\neg D)P(\neg D),$$

where $\neg D$ means not having the disease.

Now we have

- $P(+|D)$ is the sensitivity of the test;
- $P(D)$ is the prevalence;
- $P(+|\neg D) = \frac{FP}{FP+TN} = \frac{FP+TN}{FP+TN} + \frac{FP-FP-TN}{FP+TN} = 1 - \frac{TN}{FP+TN}$, which is 1 minus the specificity of the test;
- $P(\neg D)$ is 1 minus the prevalence of the test.

Therefore

$$PPV = \frac{d \times Se}{d \times Se + (1-d) \times (1-Sp)}.$$

Table 18.1   Updated contingency table, taking into account the prevalence d of a condition.

		Test result	
		Positive	Negative
Condition	Present	$d \times Se$	$d \times (1 - Se)$
	Absent	$(1 - d) \times (1 - Sp)$	$(1 - d) \times Sp$

The NPV can be calculated in the same way to

$$NPV = \frac{(1 - d) \times Sp}{(1 - d) \times Sp + d \times (1 - Se)}.$$

In the same way as sensitivity and specificity can be calculated from contingency tables, as shown in Table 8.4, PPV and NPV can also be calculated from an updated contingency table that includes prevalence d, shown in Table 18.1.

Example   Consider two biomarker based tests T_A and T_B, both achieving 85 % sensitivity and 95 % specificity. T_A is aimed at detecting a condition A that has a prevalence of 1 in 20 ($d_A = 5\%$), whereas T_B detects condition B that has a prevalence of 1 in 200 ($d_B = 0.5\,\%$).

This leads to the two updated contingency tables in Tables 18.2 and 18.3.

We can also compute the positive and negative predictive values, yielding $PPV_A = 0.472$, $PPV_B = 0.079$, $NPV_A = 0.992$, and $NPV_B = 0.999$. It is thus clear that the prevalence dramatically impacts the positive predictive value, with only 8 % of positive tests associated with the presence of the condition when the prevalence is 0.005 %; clearly an undesirably high occurrence rate of false alarms. As a general rule, conditions with low prevalence are hard to detect without yielding many false positives.

Often, the low prevalence of a disease in the general population can be addressed by first performing an orthogonal *a priori* triage. Genetic screening, presence of lumps or nodules, increased inflammation markers, and other methods can thus be

Table 18.2   Updated contingency table for test T_A applied to condition A.

		T_A result	
		Positive	Negative
Condition A	Present	0.0425	0.0075
	Absent	0.0475	0.9025

Table 18.3   Updated contingency table for test T_B applied to condition B.

| | | T_B result | |
		Positive	Negative
Condition B	Present	0.00425	0.00075
	Absent	0.04975	0.94525

used to select a first subset of the population with much increased prevalence of the condition, where the biomarker can then be employed to perform the final assessment more reliably.

△

18.2   Evaluating threshold values for biomarkers

As shown in Figure 18.1, a biomarker can show statistically significant differences in mean values between a patient and a control population, without actually being able to perfectly separate the two populations at any threshold value. Such situations are the norm rather than the exception, and it is therefore important to carefully consider the optimal threshold value for the biomarker.

One could choose to label patients by setting a low maximal value for the biomarker in this case, which means that the number of false positives will be limited (few control samples would be categorized as patients) but it would simultaneously generate many false negatives (patients incorrectly deemed controls). Conversely, setting a high maximal value will limit false negatives while yielding many false positives. It is therefore important to carefully consider the threshold to apply in light of the desired TP, FP, FN, TN prevalence and PPV and NPV.

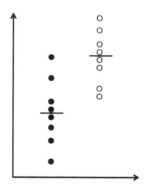

Figure 18.1   Measured values for a biomarker in patient (left-hand series of filled circles) and control samples (right-hand series of open circles), with the mean indicated by a horizontal line in each population.

18.3   Exercises

1. Derive the expression for NPV.

18.4   Bibliographic notes

Biomarker discovery   LaBaer (2005).

Pitfalls of biomarker discovery   Issaq et al. (2011).

Data processing for MS based biomarker discovery   Roy et al. (2011).

Strategies for MS based biomarker discovery and validation   Tambor et al. (2009).

19

Standards and databases

Several algorithms exist to handle quantitative data, and many of these algorithms are implemented by more than one tool. There is therefore a considerable amount of choice available with regards to processing quantitative data. However, since each software tool tends to have its own specific input and output format, it is not straightforward to swap tools, or run several tools in parallel on the same dataset. The latter has however been shown to be quite beneficial in increasing the coverage of quantifiable proteins.

Coupled with the fact that data sharing has now become common practice in the field of proteomics, it is clear that the field would benefit from standard data formats to capture the input and output of typical quantitative projects, and from centralized databases that collate and disseminate these data to the scientific community. Such standards would make it straightforward to compare, extend, or correlate results of an experiment with what is already known from previous experiments. They also allow existing datasets to be found easily for use in novel algorithm development or testing.

19.1 Standard data formats for (quantitative) proteomics

Over the past few years, several complementary data standards have been developed in proteomics by the Human Proteome Organization (HUPO) Proteomics Standards Initiative (PSI). Notable examples include the Molecular Interactions (PSI-MI) standard for protein-protein interactions, the mzML standard for mass spectrometry data, and the mzIdentML standard for mass spectrometry derived identifications. Of particular interest for the planning and communication of SRM experiments (see Chapter 15) is the TraML standard that is aimed at capturing SRM transitions and associated information.

Computational and Statistical Methods for Protein Quantification by Mass Spectrometry,
First Edition. Ingvar Eidhammer, Harald Barsnes, Geir Egil Eide and Lennart Martens.
© 2013 John Wiley & Sons, Ltd. Published 2013 by John Wiley & Sons, Ltd.

All these standards share a common design, whereby a text-based XML file is used as the framework for holding the relevant data and metadata for the format. While the actual data is often written or encoded directly into the corresponding XML tags, metadata is usually communicated through controlled vocabularies.

19.1.1   Controlled vocabularies (CVs)

A controlled vocabulary (CV) is a dictionary-like list of allowed terms, in which each term is uniquely identified by an accession number. Such a term has a preferred name (for instance, 'centroid spectrum' for the term with accession number 'MS:1000127' in the PSI-MS CV) but often synonyms are also annotated and allowed (for instance, 'Discrete Mass Spectrum' is an exact synonym for 'centroid spectrum').

Furthermore, terms can be related to each other through a set number of possible relationships. The most commonly used relationships are 'IS A,' 'HAS A,' and 'PART OF.' The term 'centroid spectrum' for instance, 'IS A' 'spectrum representation' (accession number MS:1000525), that is in turn a 'PART OF' 'spectrum' (accession number 'MS:1000442'). Other parts of a spectrum include 'spectrum attribute' (MS:1000499) and 'peak' (MS:1000231), amongst others.

As an example of how such CV terms are used in the XML document for metadata provision, take for instance the specification of the quantification method used in a hypothetical XML-based quantification standard. The 'quantification method' tag in the XML document would then take as its value any term that is a child of the 'quantification method' term (with accession number 'Q:10101') in the CV. This term has two child terms: 'MS1 based method' (Q:10206) and 'MS2 based method' (Q:10116). The latter then has a few child terms of its own, including 'Reporter ion based method' (Q:10365), that again has two child terms: 'iTRAQ' (Q:10192) and 'TMT' (Q:10800). In this example, both 'iTRAQ' and 'TMT' would therefore be applicable terms to use in the 'quantification method' tag in the XML standard, since they both are children of the 'quantification method' term in the CV. Note that it does not matter how many intermediate terms are located in between these two terms.

19.1.2   Benefits of using CV terms to annotate metadata

There are three distinct benefits to using such CV terms to annotate metadata.

Flexibility

Flexibility is achieved, since annotations can evolve along with developments in the field by simply updating the corresponding CV. In the case of our example above, consider the development of a new method for reporter based quantification in the field. It then suffices to add a new CV term as a child of the existing 'Reporter ion based method' (Q:10365) term to update the standard. Indeed, this new term would immediately qualify for inclusion in the XML document at the 'quantification method' tag, since it is indeed a child of the 'quantification method' term. No changes to the XML schema are therefore required to update the standard to include this newly developed method.

Accuracy

Use of CV terms implies readily perceived accuracy of the reported information, along with any implicit context. When a given term accession number is specified, any possible ambiguity is instantly removed, providing a very precise and unique pointer to a rigidly defined CV term. Furthermore, this term is put in a broader context through its relationships with other terms. If we for instance encounter the term 'iTRAQ' (Q:10192), we can readily see from its relationships that it 'IS A' 'Reporter ion based method' (Q:10365), which 'IS A' 'MS2 based method' (Q:10116). This allows us to conclude that the iTRAQ method is an MS2 based method that uses reporter ions, telling us a lot of useful context information about this term.

Semantic validation

The relationships provide a hierarchy that can be exploited in the semantic validation of the metadata in such an XML file. Semantic validation goes beyond syntactic validation in that the latter only verifies whether the document is correctly formatted, ignoring the actual information specified, while the former does verify whether the specified information makes sense. We could for instance populate the 'quantification method' tag of a hypothetical quantification standard document with the term 'normalization' (Q:10059), and the file would still be syntactically correct, but it will unfortunately not make any sense. Semantic validation solves this problem by flagging the inclusion of the 'normalization' term at the 'quantification method' tag as incorrect.

The validation process is typically encoded in rules that link the different tags in the XML document with various CV terms. Usually, the link is made to a parent term, for which all children are considered correct entries. Automated semantic validation has been implemented for the HUPO-PSI standards, rendering it possible to verify the syntax as well as the contents of any such file.

19.1.3   A standard for quantitative proteomics data

At the time of writing, there is not yet a formal standard defined for quantitative proteomics data, although the HUPO PSI working group on proteomics informatics (PSI-PI) is at work developing a standard, tentatively called mzQuantML. The finalization of this standard is likely to take some more time, since all PSI standards need to proceed through a peer-review process that includes a period during which the standard is open for public comments, as well as invited expert comments. However, once finalized, mzQuantML is likely to become quite popular, especially since no other consensus standard has yet emerged.

An intermediate solution is found by developing a simpler ad hoc format to capture quantitative information at lower resolution but at greater ease of development and implementation. This rather simple yet usable standard is the mzTab format, developed in the form of a tab-delimited text file. mzTab can already be used to deposit data in the popular Proteomics Identifications (PRIDE) Database at the European Bioinformatics Institute, making it a functional and supported standard to fill the vacuum before mzQuantML is released.

19.1.4 HUPO PSI

The HUPO PSI web page has a lot of information about the various proteomics data standards that have been, or are being, developed, as well as about the formal document process that each standard needs to complete. The Ontology Lookup Service (OLS) at the European Bioinformatics Institute provides a very convenient and simple way to explore controlled vocabularies and ontologies, including those from HUPO PSI.

19.2 Databases for proteomics data

Various databases and repositories exist for proteomics data, and these differ in their scope, service level, and intended applications. Repositories such as the PRIDE database at the European Bioinformatics Institute specialize in capturing and maintaining publication-related proteomics data, with the main focus on supporting data submission and subsequent retrieval. Data submission to PRIDE is greatly facilitated by the powerful yet user friendly PRIDE Converter tool, that can transform a variety of input formats into PRIDE XML that can be directly uploaded to PRIDE. A particularly interesting PRIDE-associated feature is the Proteomics Identifier Cross-Referencing (PICR) service, that can be used to translate protein accession numbers from one type into another, even taken into account changes over time.

Research oriented databases also exist, where the storage of proteomics data is primarily meant to support research into novel computational tools for mass spectrometry based proteomics. Notable examples of this category are PeptideAtlas and GPMDB. These two systems are quite comparable in concept and execution, with both databases serving as the endpoint of a fixed and uniform processing pipeline that takes the original spectral data as input to produce identifications. Both systems have also led to the same type of analyses and novel applications, including the identification of proteotypic peptides as well as spectral library searching algorithms. None of these systems is particularly adept at storing or disseminating quantitative data, and only the PRIDE repository currently makes any attempt at formalizing data submission, storage and dissemination for quantitative data sets through the above mentioned mzTab format. It is expected that these databases will however start capturing quantitative data after the mzQuantML standard has been released, and has gained sufficient traction in the field.

A large amount of protein quantification data across different tissues and even between tumor versus normal tissue has however been obtained and cataloged by the Protein Atlas project. The quantification is here based on protein-specific antibody reactivity rather than mass spectrometry. Nevertheless, this resource can be of use if comparisons are sought between observed and previously known protein quantities.

19.3   Bibliographic notes

Processing quantitative data

Mueller et al. (2008), Vaudel et al. (2010), Colaert et al. (2011b).

Standards

- PSI-MI: Hermjakob et al. (2004);
- mzML: Martens et al. (2011);
- mzIdentML: Jones et al. (2012);
- TraML: Deutsch et al. (2012);
- mzQuantML: Vizcano et al. (2007).

Semantic validation

Montecchi-Palazzi et al. (2009).

Databases

- PRIDE: Martens et al. (2005), Barsnes et al. (2009), Côtè et al. (2007);
- PeptideAtlas: Desiere et al. (2006);
- GPMDB: Craig et al. (2004), Craig et al. (2005), Craig et al. (2006), Lam et al. (2007);
- Protein Atlas: Nilsson et al. (2005).

Web addresses for standards, controlled vocabularies and accession number conversion

- HUPO PSI (www.psidev.info);
- The Ontology Lookup Service (OLS; www.ebi.ac.uk/ols);
- The Protein Identifier Cross-Referencing Service (PICR; www.ebi.ac.uk/Tools/picr).

Web pages for databases

- PRIDE (www.ebi.ac.uk/pride);
- PeptideAtlas (www.peptideatlas.org);
- GPMDB (www.gpmdb.org);
- Protein Atlas (www.proteinatlas.org).

20

Appendix A: Statistics

Statistics is an important part of protein quantification projects. Though the statistical analysis can be done by statisticians, the other participants in the project should have a basic understanding of it. In this appendix we describe the essential parts of statistics related to quantification projects. It is, however, not an introduction to statistics, of which the readers are assumed to have a basic knowledge. The goal here is to collect a brief explanation of the most used concepts in a compact and unified way.

20.1 Samples, populations, and statistics

Statistics can be defined as a scientific field consisting of four subfields:

- collection of data;

- analysis of data;

- interpretation or explanation of data;

- presentation of data.

The data analyzed in quantitative protein projects usually constitute *samples* from *statistical populations*. In this book we will use the term 'population' for statistical populations when it cannot be misunderstood. A sample is often defined by a set of observed values of a random variable X. The values are sampled from a population of values, on which there often exists a *probability distribution*. A probability distribution defines:

- the probability of X having the value x ($P(X = x)$) if X is discrete;

- the probability of X falling in a specific interval if X is continuous, for example, the probability that X is greater than x ($P(X > x)$).

Computational and Statistical Methods for Protein Quantification by Mass Spectrometry,
First Edition. Ingvar Eidhammer, Harald Barsnes, Geir Egil Eide and Lennart Martens.
© 2013 John Wiley & Sons, Ltd. Published 2013 by John Wiley & Sons, Ltd.

For a discrete random variable the probability distribution is given by a formula for the point probabilities, and for a continuous random variable by a formula for the probability density. Both types of distributions can be specified by a cumulative distribution function. Often the probability distribution is specified except for some unknown parameters, for example, the mean and the standard deviation in a normal distribution. To draw inference about the unknown parameters, relevant data is often collected in the form of a *random sample* from a defined population.

A random sample of size n is a vector of n random variables $X = (X_1, X_2, \ldots, X_n)$. And a *statistic* (singular) is a random variable that is a *function* of a sample. When a set of values for the n random variables are observed we say that a sample is taken and the observed value of the statistic can be calculated, for example, the sample mean or the sample variance, or different forms of *test statistics*.

When a statistic is used to estimate an unknown parameter of a distribution it is called an *estimator* and the calculated value of it is an *estimate*. It is worth noting that a statistic is a random variable with a probability function determined by the distribution of the random sample. In everyday language it is common to not discriminate between the statistic (T) as a random variable and an observed value of it (t), or between the random sample and an observed value for the random sample. In statistics it is however common to denote random variables with upper case letters, and values of random variables by the corresponding lower case letters, a convention adopted in this book.

Example We want to investigate the abundance of a specific protein in humans at a given time point after the intake of a drug. n persons are given the drug. The abundance from these n persons constitute the sample, from which we can calculate the statistics sample mean and sample variance. The abundances that could be measured if all living people had gotten the drug constitute the population, of which the parameters mean and variance are unknown.

$\triangle$

20.2   Population parameter estimation

Generally there are two ways of estimating a population parameter ϕ from a sample $X = (X_1, X_2, \ldots, X_n)$, *point estimation* and *confidence interval estimation*.

Point estimation   means that a value for ϕ is calculated from a value for X. Depending on the size of X the probability distribution of the population must also be taken into account.

Confidence interval   is used to include an accuracy of the estimation. A *confidence level* is defined, usually in percent c, for example, 95 % confidence level, and the interval is called a c % confidence interval. Confidence intervals are typically calculated by first defining a probability α (e.g., 0.05), and then from a sample calculating the *confidence limits* $l(X), h(X)$, such that

$$P(l(X) < \phi < h(X)) = 1 - \alpha \qquad (20.1)$$

The interval $(l(X), h(X))$ is called a $100(1 - \alpha)\%$ confidence interval for ϕ with confidence level $100(1 - \alpha)\%$ (or $1 - \alpha$). This says that *when a large number of random samples are taken, and values for the confidence interval is calculated for each sample, approximately $100(1 - \alpha)\%$ of the intervals will cover ϕ.* It is important to note that it *does not say that it is $100(1 - \alpha)\%$ probability that ϕ is in a specific interval $(l(x), h(x))$.*

We mentioned that the estimations can depend on the distribution of the population. Often one assumes a normal, or nearly normal, distribution. Remember that the normal distribution has the density function

$$\frac{1}{\sqrt{2\pi}\,\sigma} e^{-\frac{(x-\mu)^2}{2\sigma^2}},$$

where μ, σ are the parameters' mean and standard deviation (*location* and *scale* parameter).

The normal distribution is denoted by $N(\mu, \sigma)$ and is bell-like, meaning

- it is symmetrical, hence the mean and the median are the same;
- it is unimodal, with the mode equal to the mean and median.

It is worth noting that there is nothing normal about the 'normal distribution,' and many prefer to term it the Gaussian distribution.

20.2.1   Estimating the mean of a population

As an example let us estimate the mean of a population from which a sample X is drawn, with size n, mean $\bar{X}$, and standard deviation S. If the population is normally distributed (or n is large) then the sample mean $\bar{X}$ is exactly (or approximately, from the *central limit theorem*) normally distributed $N(\mu, \frac{\sigma}{\sqrt{n}})$. This means that $\bar{X}$ is an unbiased (point) estimator for μ.

The standard deviation of $\bar{X}$ is $\frac{\sigma}{\sqrt{n}}$, and it is called *standard error of the mean* (SEM). Since σ is most often unknown, the standard error of the mean is estimated by $S/\sqrt{n}$. When n is large (> 30) the effect of using the estimated value can be neglected, and $S/\sqrt{n}$ is also (informally) often denoted as the *standard error*.

To calculate a confidence interval for μ, let Z be a standard normally distributed variable, $N(0, 1)$ and remember that z_α is defined as $P(Z \geq z_\alpha) = \alpha$. Then

$$P(-z_{\alpha/2} < Z < z_{\alpha/2}) = 1 - \alpha.$$

Let Y be normally distributed $N(\mu, \sigma)$. Then $\frac{Y-\mu}{\sigma}$ is standard normally distributed, so

$$P\left(-z_{\alpha/2} < \frac{Y-\mu}{\sigma} < z_{\alpha/2}\right) = 1 - \alpha$$

and

$$P\left(\mu - z_{\alpha/2}\sigma < Y < \mu + z_{\alpha/2}\sigma\right) = 1 - \alpha. \tag{20.2}$$

Let X be a random sample such that $\bar{X}$ is (exact or approximately) normally distributed $N(\mu, \frac{\sigma}{\sqrt{n}})$. From Equation 20.2 we achieve

$$P\left(\mu - z_{\alpha/2}\frac{\sigma}{\sqrt{n}} < \bar{X} < \mu + z_{\alpha/2}\frac{\sigma}{\sqrt{n}}\right) = 1 - \alpha.$$

Hence a $100(1 - \alpha)\%$ confidence interval for μ is

$$\left(\bar{X} - z_{\alpha/2}\frac{\sigma}{\sqrt{n}}, \bar{X} + z_{\alpha/2}\frac{\sigma}{\sqrt{n}}\right). \tag{20.3}$$

Example Let the standard deviation of a population be $\sigma = 16$, and choose $\alpha = 0.05$. From the standard normal distribution table we have $P(Z < 1.96) = 0.975$, hence $z_{\alpha/2} = 1.96$. Let a sample of size $n = 64$ be randomly drawn from this population, with the mean $\bar{x} = 20$. Then the 95 % confidence interval for μ becomes $(16.08, 23.92)$. Remember that this *does not say that it is a probability of 0.95 that μ will be in this interval.*
△

Usually σ is unknown, but can be replaced by the sample standard deviation without changing the effect when the sample is large. If $\bar{X}$ is normally distributed and n is small the z-values from the normal distribution should be replaced by the corresponding values from a T distribution with $n - 1$ degrees of freedom giving slightly wider intervals for decreasing n.

20.3   Hypothesis testing

Hypothesis testing can briefly be described as follows:

1. Usually two mutually exclusive and extensive competing hypotheses are formulated. The *null hypothesis* H_0 is the hypothesis that we want to reject, usually an hypothesis of no difference or no effect. The *alternative hypothesis*, H_A, is usually the hypothesis we want to prove indicating that there is a difference or effect. We then decide on the risk we are willing to except for falsely concluding that H_A is correct, the *significance level*, α. It is common to let α be 0.05, but with multiple testing a lower level has to be chosen for each test.

2. We perform a relevant experiment to test H_0 and record values of a random sample from which we calculate the value of a test statistic.

3. Under the assumption that H_0 is the correct hypothesis we calculate the probability, p, of observing a result at least as extreme as what we have observed. This is the *significance probability* or *p-value*. If $p \leq \alpha$ H_0 is *rejected* and H_A *accepted*.

20.3.1   Two types of errors

When performing hypothesis testing two types of errors can occur.

Type I error   means rejecting H_0 when H_0 is true. The probability of a Type I error is the significance level α. Type I error is also called the *rejection error*.

Type II error   means accepting H_0 when H_0 is false. The probability of a Type II error is often denoted by β. Type II error is also called the *acceptance error*.

The *power* of a statistical test is defined as the probability that it will reject H_0 when H_0 is false. This means the probability that the test will not make an error of Type II, that is, the power is $1 - \beta$. The power is also sometimes denoted *sensitivity*.

20.4   Performing the test – test statistics and p-values

To test the null hypothesis a *random variable V* is used of which a value can be calculated from the measured data (samples) under the assumption that H_0 is valid. Let such a calculated value be v_0. One then finds the probability of achieving v_0 or more extreme values when H_0 is valid.[1] This probability is the p-value associated with v_0, and it is compared to α for determining rejection or not.

The random variable V is called a *test statistic*. For most of the common test statistics the exact function is complicated. However, statistical computer programs exist that easily calculate the value of V and the corresponding p-values for a variety of statistical tests.

Example   Consider the task of investigating if a protein P is differentially abundant in two situations. H_0 is that they are not differentially abundant, and let $\alpha = 0.05$. We perform five experiments for each situation, thus achieving five values for the abundance of P in each situation. From these we calculate values for the mean and variance for each situation, and further calculate the value of a T-statistic (using Equation 20.5 in Section 20.5.2) to $t_0 = 2.602$.

By using a computer program (or looking in a table) we find the probability of observing a value for the T-statistic with 8 degrees of freedom which is at least 2.602

---

[1] More extreme can be higher, less, or higher or less in absolute value depending on the test.

in absolute value (the p-value) to be 0.0315 which is less than 0.05. We therefore reject H_0 and conclude that P is differentially abundant in the two situations.
△

We see that a test statistic is a measure of the conformity between a null hypothesis and the collected (sample) data. Thus a test statistic is generally a function of the sample and the population parameters that are to be tested, $V = v(X, \phi)$.

When it is not possible (or difficult) to use any of the well-known test statistics for which the distributions are known, one has to define one's own statistics, and then in some way determine a distribution for them.

Example We have two protein sequences d, q, and want to test if they are homologous. The null hypothesis is that they are not. We then align the sequences, and find the score s_1 of the highest scoring local alignment, based on some scoring scheme. The scores s can be considered as a test statistic, and we must construct a probability distribution for the score when nonhomologous sequences (of the same form as q and d) are aligned.
△

The test statistics can be divided into two types, depending on the assumptions one makes regarding the underlaying population.

20.4.1   Parametric test statistics

Parametric test statistics are test statistics that can be used when the sample comes from a defined probability distribution, for example, from populations that are normally distributed. In our quantification project it means that the measured values for each situation constitute a sample from a normally distributed population. It is called parametric since the *parameters* of the underlaying populations are used, often estimated from the sample, for example, mean and standard deviation. This means that when we use parametric statistics we make assumptions about the population from which the sample is collected. Examples of such statistics are Z-statistics, T-statistics, F-statistics, G-statistics, and Chi-square statistics.

Many test statistics require that the populations from where the data is taken are normally, or nearly normally, distributed, at least for small samples. In addition, the individual observations must be mutually independent of each other.

20.4.2   Nonparametric test statistics

Nonparametric statistical methods do not require that the sample is taken from any defined probability distribution. They are said to be more *robust* than parametric statistics, since they make fewer assumptions about data that is to be analyzed.

If the original sample $(X_1, X_2, \ldots, X_n)$ is rearranged so that its values are put in ascending order, we sometimes call the ordered sample, $(X_{(1)}, X_{(2)}, \ldots, X_{(n)})$, the *order statistic*. Various other statistics can be based on the ordered sample, for instance the *range*, $X_{(n)} - X_{(1)}$, and the *median*, $X_{((n+1)/2)}$ if n is odd and $(X_{(n/2)} + X_{(n/2+1)})/2$ if n is even.

Certain tests are based on the *ranks* $(R_1, R_2, \ldots, R_n)$ of the sample. The rank R_i of the observation X_i equals X_i's position in the order statistic. Often a nonparametric statistic is defined by just replacing the original sample by the ranks of the sample.

Example Let the observed sample be $(0.6, 1.7, -0.5, 0.3, 1.4)$. Then the ordered sample becomes $(-0.5, 0.3, 0.6, 1.4, 1.7)$ and the corresponding ranks are $(3, 5, 1, 2, 4)$. The range is $1.7-(-0.5) = 2.2$ and the median is 0.6.
$\triangle$

20.4.3   Confidence intervals and hypothesis testing

Confidence intervals are also used for hypothesis testing. Assume that we have a population with an unknown parameter ϕ and want to test $H_0 : \phi = \phi_0$. From an observed sample suppose we can calculate values for a $100(1 - \alpha)\,\%$ confidence interval. Then we reject H_0 at significance level α if the calculated confidence interval covers ϕ.

Example Let us test the population mean $\mu = \mu_0$. Then for large n the Z-statistic $Z = \frac{\bar{X}-\mu_0}{S/\sqrt{n}}$ can be used. Note that in Section 20.2.1 we performed the test assuming that the population standard deviation was known. Here we use S as an estimator for σ.

From $P(Z \geq z_{\alpha/2}) = \frac{\alpha}{2}$ we get

$$P\left(\frac{\bar{X} - \mu}{S/\sqrt{n}} \geq z_{\alpha/2}\right) = \frac{\alpha}{2}$$

and

$$P\left(\bar{X} - z_{\alpha/2}\frac{S}{\sqrt{n}} \geq \mu\right) = \frac{\alpha}{2}.$$

Since the Z-statistic is symmetric we derive the following:

$$P\left(\bar{X} - z_{\alpha/2}\frac{S}{\sqrt{n}} < \mu < \bar{X} + z_{\alpha/2}\frac{S}{\sqrt{n}}\right) = 1 - \alpha.$$

Thus a $100(1 - \alpha)\%$ confidence interval is $(\bar{X} - z_{\alpha/2}\frac{S}{\sqrt{n}}, \bar{X} + z_{\alpha/2}\frac{S}{\sqrt{n}})$, and H_0 is rejected if

$$\bar{X} - z_{\alpha/2}\frac{S}{\sqrt{n}} < \mu_0 < \bar{X} + z_{\alpha/2}\frac{S}{\sqrt{n}}$$

is not satisfied (μ_0 is outside the confidence interval).
$\triangle$

We see that from a confidence interval we can test a set of hypotheses, each having different μ_0. We also see from the understanding of the confidence interval that for a set of tests for which H_0 is valid approximately 100α % of them will reject H_0, corresponding to a probability of Type I error of α.

20.5   Comparing means of populations

A common task in proteomics is to compare the means of two or more populations, for example, the abundances of proteins in cod living in clean water versus cod living in polluted water.

When presenting the statistics used it is common to distinguish between the case where there are exactly two populations in the analysis, and the case where there are more than two. The reason for this is that when we have two populations we can employ simpler statistics.

When comparing the means between populations we should have the following in mind:

- the *distributions* of the populations must be considered, since this determines which statistics that can be used;

- since the parameters of the populations in most cases are unknown, we must estimate them from samples;

- how the samples are drawn from the populations is important.

The null hypothesis we are testing is most often that there are no difference, or also for two populations that the difference is equal to a constant k.

To describe the statistics we start by considering the mean of a single population.

20.5.1   Analyzing the mean of a single population

For the mean of a population the null hypothesis is generally $H_0 : \mu = \mu_0$.

- for large n the Z-statistic can be used $Z = \frac{\bar{X} - \mu_0}{S/\sqrt{n}}$;

- for small n and normally distributed populations the T-statistic can be used $T = \frac{\bar{X} - \mu_0}{S/\sqrt{n}}$.

The formula for the two statistics are equal, but the probability distributions for them are different. The Z-statistic is standard normally distributed, and the distribution for the T-statistic is called a T *distribution* or *Student's T* distribution. There are different T distributions depending on the sample size n, characterized by the number of *degrees of freedom*, or *df*. In this case the degree of freedom is $n - 1$.

The test for $\mu \neq \mu_0$ is then performed by a computer program calculating the p-value, that is, $P(|T| > v(\alpha/2))$ where v is either from a normal distribution or a T distribution with $n - 1$ degrees of freedom.

Example Suppose $n = 25$, $\bar{X} = 0.6$, $S = 2.2$. H_0 is $\mu_0 = 0$, and $\alpha = 0.05$. A computer program calculates $p = 0.1738$ using the normal distribution and $p = 0.1865$ using the T distribution with 24 degrees of freedom. H_0 is therefore not rejected.
$\triangle$

Confidence interval

In Section 20.4.3 we calculated a confidence interval when the Z-statistic was used. Since the T-statistic has the same formula, we achieve the $(1 - \alpha)100\%$ confidence interval as:

$$\left(\bar{X} - t_{\alpha/2}\frac{S}{\sqrt{n}}, \bar{X} + t_{\alpha/2}\frac{S}{\sqrt{n}} \right),$$

where the degree of freedom for T is $n - 1$.

Example Assume that we have measured the abundances of a protein P from liver in five 1-year-old cod living in clean water to be 3000, 3000, 4000, 5000, 5000 respectively. Then $\bar{X} = 4$, and $S = 1$. To calculate the 95 % confidence interval we find $t_{0.025}(4) = 2.776$. The estimated standard error of the mean is $SEM = \frac{S}{\sqrt{n}} = 0.447$. Thus the 95 % confidence interval is $4 \pm 2.776 \cdot 0.447 = 4 \pm 1.24 = (2.76, 5.24)$. We say that we have 95 % confidence that the interval $(2.76, 5.24)$ covers the true population mean.
$\triangle$

20.5.2   Comparing the means from two populations

The general hypothesis here is $H_0 : \mu_1 - \mu_2 = \delta_0$, where δ_0 is often zero. Assume

- two normally distributed populations $N(\mu_1, \sigma_1)$ and $N(\mu_2, \sigma_2)$;
- a sample is to be randomly and independently drawn from each of the populations with sizes, means, and standard deviations n_1, $\bar{X}_1$, S_1 and n_2, $\bar{X}_2$, S_2.

We then have the following:

- $\bar{X}_1 - \bar{X}_2$ is normally distributed $N(\mu_1 - \mu_2, \sqrt{\frac{\sigma_1^2}{n_1} + \frac{\sigma_2^2}{n_2}})$.
- It then follows that

$$Z = \frac{(\bar{X}_1 - \bar{X}_2) - (\mu_1 - \mu_2)}{\sqrt{\frac{\sigma_1^2}{n_1} + \frac{\sigma_2^2}{n_2}}} \qquad (20.4)$$

is standard normally distributed.

- Since the population standard deviations generally are unknown, we should again estimate them from the sample variances. However, the formulas used for estimation are different depending on the population variances being equal or different.

Assuming equal population variances

When the two population variances are equal to σ^2, we get from Equation 20.4:

$$Z = \frac{(\bar{X}_1 - \bar{X}_2) - (\mu_1 - \mu_2)}{\sigma\sqrt{\frac{1}{n_1} + \frac{1}{n_2}}}.$$

σ should now be estimated by using the sample standard deviations. These are probably different and it seems reasonable to trust highest the standard deviation calculated from the largest sample. This can be taken care of by a *pooled* estimation

$$S_p^2 = \frac{(n_1 - 1)S_1^2 + (n_2 - 1)S_2^2}{n_1 + n_2 - 2}.$$

Using this we get that

$$T = \frac{(\bar{X}_1 - \bar{X}_2) - (\mu_1 - \mu_2)}{S_p\sqrt{\frac{1}{n_1} + \frac{1}{n_2}}}$$

follows a T distribution with $n_1 + n_2 - 2$ degrees of freedom. This can be used to test $H_0 : \mu_1 - \mu_2 = \delta_0$ by the *two-sample T-statistic with $df = n_1 + n_2 - 2$*:

$$T = \frac{(\bar{X}_1 - \bar{X}_2) - \delta_0}{S_p\sqrt{\frac{1}{n_1} + \frac{1}{n_2}}}. \tag{20.5}$$

From Equation 20.5 we can now derive a $100(1 - \alpha)\%$ confidence interval for $\mu_1 - \mu_2$ as:

$$(\bar{X}_1 - \bar{X}_2) \pm t_{\alpha/2} S_p\sqrt{\frac{1}{n_1} + \frac{1}{n_2}}, \tag{20.6}$$

where $df = n_1 + n_2 - 2$ for $t_{\alpha/2}$.

Assuming unequal population variances

When the two population variances are unequal, we must estimate each of them with their sample variances, hence the two-sample T-statistic now becomes:

$$T = \frac{(\bar{X}_1 - \bar{X}_2) - \delta_0}{\sqrt{\frac{S_1^2}{n_1} + \frac{S_2^2}{n_2}}}. \tag{20.7}$$

The drawback with this is that it does not follow a T distribution. However, it can be approximated by calculating an approximation to the degrees of freedom. For easy use the minimum of n_1 and n_2 can be used as a pessimistic value. A more exact value is defined by Welch, referred to as *Welch's T-test*, where the degree of freedom is approximated as:

$$W = \frac{(\frac{S_1^2}{n_1} + \frac{S_2^2}{n_2})^2}{\frac{S_1^4}{n_1^2(n_1-1)} + \frac{S_2^4}{n_2^2(n_2-1)}}. \tag{20.8}$$

This is a sufficient approximation when the sample sizes are at least five.

Large samples

When the samples are large (> 30) the Student's T distribution approaches the standard normal distribution. This is also valid even if the populations are not normally distributed (from the central limit theorem), and we can derive that

$$\frac{(\bar{X}_1 - \bar{X}_2) - (\mu_1 - \mu_2)}{\sqrt{\frac{S_1^2}{n_1} + \frac{S_2^2}{n_2}}}$$

has approximately a standard normally distribution. Note that this is also valid for σ_1 different from σ_2.

Thus the Z-statistic for $H_0 : \mu_1 - \mu_2 = d_0$ becomes:

$$Z = \frac{(\bar{X}_1 - \bar{X}_2) - \delta_0}{\sqrt{\frac{S_1^2}{n_1} + \frac{S_2^2}{n_2}}}.$$

In the same way as before we can derive an approximation to the $100(1 - \alpha)\%$ confidence interval for $\mu_1 - \mu_2$ as:

$$(\bar{X}_1 - \bar{X}_2) \pm z_{\alpha/2}\sqrt{\frac{S_1^2}{n_1} + \frac{S_2^2}{n_2}}. \tag{20.9}$$

20.5.3 Comparing means of paired populations

In some experiments it may be appropriate to pair the values from the two (statistical) populations. One example can be investigating the effect of a specific drug, and a pair is then the protein abundances of the same person before and after he/she has taken the drug.

Let the two samples be paired so that X_i is paired to Y_i for all persons, and assume all pairs are independent. Define $D_i = X_i - Y_i$, and let $\bar{D}$ and S_D be the mean and standard deviation of the sample differences. Assume that the D_is are normally distributed, and define δ to be the difference between the two population means. For testing δ we can then derive:

- Under $H_0 : \delta = \delta_0$, the test-statistic

$$T = \frac{\bar{D} - \delta_0}{S_D/\sqrt{n}}$$

  is T-distributed with $df = n - 1$ and can be used for testing H_0.

- A $100(1 - \alpha)$ % confidence interval for δ is

$$\bar{D} \pm t_{\alpha/2} \frac{S_D}{\sqrt{n}}.$$

Robustness of the T-statistic

To be able to use the T-statistic on small samples we have assumed approximate normally distributed populations. The larger the samples are, the less effect it will have that the populations are not normally distributed. When the sample size is over 15, and the population distribution is not too skewed, the effect seems to be small. Following this we can say that the T-statistic is fairly robust for drawing inferences about μ.

20.5.4 Multiple populations

When the means of several groups shall be compared one can pair-wise compare all of the possible pairs, using the methods as explained above. This is cumbersome, and the overall risk of Type I error is inflated if no specific precautions are taken. However, it is possible to compare all means in one test, but note that this does not give the same type of results as the pair-wise approach.

ANOVA is the most used method, which is an acronym for ANalysis Of VAriance. ANOVA partitions the observed variance in a random variable into components attributable to different sources of variation, and the name follows from this. The main purpose of doing an ANOVA rather than multiple T-tests is to control the overall probability of making a Type I error.

The question answered by ANOVA is *do all groups have the same population mean?* This means that if the answer is no, it cannot tell us which means are different.

To be able to use the ANOVA procedure one has to assume that all population variances are equal, but the sample sizes can vary. Under the assumption of normality or approximate normality the result of an ANOVA is given in terms of an F-test. Without the normality assumption a nonparametric ANOVA can be performed. An ANOVA can also be performed as a special case of a linear regression analysis.

20.5.5   Multiple testing

Often one simultaneously performs several tests, and the chances of performing errors of Type I or II for at least one of them increases with the number of tests. This is important when trying to identify differently abundant proteins in a sample of proteins. Multiple testing in proteomics is discussed in Chapter 8.5.

20.6   Comparing variances

We have seen that for small normally distributed populations the statistics used for comparing means depend on the variances being equal or different. There is therefore a need for testing this. Note that the test methods described are *very sensitive to the populations being normally distributed*. We first consider the variance of a single population.

20.6.1   Testing the variance of a single population

The most used test statistic for testing the variance in this case is the *chi-square test statistic* which has a *chi-square distribution*

$$Q^2 = \frac{(n-1)S^2}{\sigma^2},$$

with $df = n - 1$.

The test statistic used for $H_0 : \sigma^2 = \sigma_0^2$ is then

$$Q^2 = \frac{(n-1)S^2}{\sigma_0^2}.$$

A chi-square distribution is not symmetric, and this must be taken into consideration when the one-sided or two-sided tests are performed.

The asymmetry of the chi-square distribution must also be taken care of when calculating a confidence interval. For a $100(1 - \alpha)\%$ confidence interval we divide α equally on the two tails getting

$$P\left(q_{(1-\alpha/2)} < \frac{(n-1)S^2}{\sigma^2} < q_{\alpha/2}\right) = 1 - \alpha.$$

From this a confidence interval for σ^2 is

$$\left(\frac{(n-1)S^2}{q_{\alpha/2}}, \frac{(n-1)S^2}{q_{1-\alpha/2}} \right),$$

with $df = n - 1$ for Q^2.

20.6.2   Testing the variances of two populations

The most used test statistic for normally distributed population when comparing variances is the F-test statistic which follows an F distribution

$$F = \frac{\frac{S_1^2}{\sigma_1^2}}{\frac{S_2^2}{\sigma_2^2}},$$

where 1 and 2 refer to the two populations. This statistic has two degrees of freedom, and for the variance case they are $(n_1 - 1, n_2 - 1)$. It is therefore common to use the notation $F(v_1, v_2)$, where v_1, v_2 are the two degrees of freedom.

The null hypothesis is that the two variances are equal, $H_0 : \sigma_1^2 = \sigma_2^2$, which is the same as

$$H_0 : \frac{\sigma_1^2}{\sigma_2^2} = 1.$$

If H_0 is true, F becomes $\frac{S_1^2}{S_2^2}$ and for the test we must consider how far this is from 1, and compare to a critical value corresponding to significance level α.

An F distribution is not symmetric, which makes the test a bit more complicated in that we must treat each end of the distribution separately.

The asymmetry means that for the test we can consider one of the three alternative hypotheses:

$$H_1 : \frac{\sigma_1^2}{\sigma_2^2} > 1, \; H_1 : \frac{\sigma_1^2}{\sigma_2^2} < 1, \; H_1 : \frac{\sigma_1^2}{\sigma_2^2} \neq 1$$

We now use the following property of the F distribution:

$$f_{1-\alpha}(n_1 - 1, n_2 - 1) = \frac{1}{f_\alpha(n_2 - 1, n_1 - 1)}$$

to derive

$$P\left(\frac{1}{f_{\alpha/2}(n_2 - 1, n_1 - 1)} < \frac{S_1^2 \, \sigma_2^2}{S_2^2 \, \sigma_1^2} < f_{\alpha/2}(n_1 - 1, n_2 - 1) \right) \tag{20.10}$$
$$= 1 - \alpha.$$

Redefine $F = (\frac{S_1}{S_2})^2$. Then the rejection of H_0 can be formulated as:

- $H_1 : \frac{\sigma_1^2}{\sigma_2^2} > 1$; reject if $F \geq f_\alpha(n_1 - 1, n_2 - 1)$;

- $H_1 : \frac{\sigma_1^2}{\sigma_2^2} < 1$; reject if $F \leq \frac{1}{f_\alpha(n_2 - 1, n_1 - 1)}$;

- $H_1 : \frac{\sigma_1^2}{\sigma_2^2} \neq 1$; reject if $F \geq f_{\alpha/2}(n_1 - 1, n_2 - 1)$ or $F \leq \frac{1}{f_{\alpha/2}(n_2 - 1, n_1 - 1)}$.

From Equation 20.10 it is easy to formulate a confidence interval for $\frac{\sigma_2^2}{\sigma_1^2}$, from which a $100(1 - \alpha)\%$ confidence interval for $\frac{\sigma_1^2}{\sigma_2^2}$ can be derived:

$$\left(F \frac{1}{f_{\alpha/2}(n_1 - 1, n_2 - 1)}, F f_{\alpha/2}(n_2 - 1, n_1 - 1) \right).$$

20.7 Percentiles and quantiles

When one has a distribution or a sample of a random variable X, one often wants to find values that divide the *ordered* values of X into two parts, for example, a division point q for which $p \%$ of the values are less than q, and $100 - p \%$ are greater than or equal to q. Such values for q are called *percentiles* or *quantiles*. To specify which % are used they are termed p-th percentile. The 50-percentile of a sample corresponds to the median.

A standard definition of percentiles does not exist. One tentative (and seemingly reasonable) definition could be a value z such that:

- $p \%$ of the values in the sample are less than z; and

- $(100 - p) \%$ are greater than z.

A problem with this definition for finite samples is that such a z does not always exist, or that a lot if them exist, that is, an interval.

Example We have a sample with ordered values (4, 6, 8, 9, 9, 11, 12, 14) and let $p = 50$. We do not find a z satisfying the definition above.
$\triangle$

To deal with this problem several definitions have been proposed. When the sample size is large they all give similar results, but for smaller sample sizes the found values for the percentile can be substantially different. Thus various statistical programs can give different values for the percentiles.

We will here describe one method for estimating the percentiles.

20.7.1   A straightforward method for estimating the percentiles

A straightforward definition can be: Given an ordered sample of n values, and a value $p : 0 < p \leq 100$, the p-th percentile is a value q such that:

- at least p % of the numbers in the series are less or equal to q; and

- at least $(100 - p)$ % are greater or equal to q.

Example   Consider the sample data (7.9, 8.3, 8.4, 9.2, 9.7, 10.3, 10.3, 10.5, 11.2, 11.3, 11.7, 12.1, 12.4, 12.6, 12.8). The number of observations $n = 15$. For $p = 30$ we require that:

- at least 4.5 of the numbers are less or equal to q;

- at least 10.5 of the numbers are greater or equal to q.

To achieve this the calculated numbers must be transformed to integers, 5 and 11. From this we find: $q \geq 9.7$ and $q \leq 9.7$, showing that the 30th percentile is 9.7.
For $p = 60$ we find:

- at least 9 of the numbers must be less or equal to q, meaning $q \geq 11.2$;

- at least 6 of the numbers must be greater or equal to q, meaning $q \leq 11.3$.

From this we see that q can be (according to the definition) any number in the interval [11.2, 11.3]. In such cases one often take the average as the percentile, hence here $q = 11.25$.
$\triangle$

Let our data in increasing order be denoted $u_1, u_2, \ldots, u_n$. Generally we can then show:

- if $np/100$ is an integer i then $q \in [u_i, u_{i+1}]$;

- if $np/100$ is not an integer then $q = u_{\lfloor np/100 \rfloor}$.

20.7.2   Quantiles

For specific values of p the percentiles are often denoted as *quantiles*. A q-quantile defines $q - 1$ values, and if we use the straightforward definition the i-th q-quantile q_i is for $0 < i < q$ defined as:

- at least i/q of the numbers are $\leq q_i$; and

- at least $(q - i)/q$ of the numbers are $\geq q_i$.

The quantiles have specific names for specific qs, for example:

- the 3-quantiles are denoted *terciles* or *tertiles*, and T is used as an abbreviation;

- the 4-quantiles are denoted *quartiles*, Q as abbreviation;

- the 5-quantiles are denoted *quintiles*, U as abbreviation;
- the 100-quantiles are denoted *percentiles*, P as abbreviation;
- the 1000-quantiles are denoted *permillages*, Pr as abbreviation.

From this we see, for example:

- the first quartile, Q_1 is the 25th percentile;
- the second quartile, Q_2 is the 50th percentile, and also the median;
- the third quartile, Q_3 is the 75th percentile.

Example Assume that we have an ordered list of ten observations 35, 53, 55, 60, 60, 65, 70, 72, 73, 78. The first quartile (Q_1) is determined as:

- at least $\lceil 10/4 \rceil = 3$ of the observations must be $\leq Q_1$, meaning $Q_1 \geq 55$;
- at least $\lceil 30/4 \rceil = 8$ of the observations must be $\geq Q_1$, meaning $Q_1 \leq 55$;
- from this $Q_1 = 55$.

Further we find $Q_2 = (60 + 65)/2 = 62.5$, $Q_3 = 72$.

$\triangle$

An advantage of percentiles and quantiles is that they are not so sensitive to outliers (Chapter 6.4).

20.7.3   Box plots

A box plot is a graphical presentation of a distribution of numbers showing five numbers:

- the largest value;
- the smallest value;
- the median;
- the quartiles Q_1 and Q_3.

Figure 20.1 shows an example. Various statistical software may use different algorithms for drawing box plots, for example, highlighting outliers and extreme outliers, and the box plots must be interpreted according to this.

20.8   Correlation

Detecting coherences, similarities, and differences in the protein profiles from different samples or experiments is an essential part of many proteomic analyses. This is most often done by performing correlation analyses and/or clustering.

Correlations are often illustrated in a *scatter plot*.

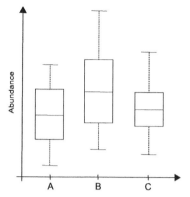

Figure 20.1   Illustration of a box plot. Q_1 for C is larger than Q_1 for A, but for Q_3 it is the opposite.

Scatter plot   A scatter plot (or scattergram), is a diagram with each of the variables represented at each axis, as shown in Figure 20.2. A scatter plot is a visual presentation of the relationship between two variables, and may give a hint if it is worth while exploring it further for a specific (e.g., linear) relation.

For a positive variable with a right-skewed distribution a log scale is often used. This may imply a log-scale for both or either of the two variables in the scatter plot. Figure 20.2 shows scatter plots of original data and transformed data.

Other examples where examining correlations can be interesting are:

- Is there a correlation of the abundances of two specific proteins in a site across different states or time periods? In such an examination the abundances of the two proteins are on the different axes. In a diagram there will be one point for

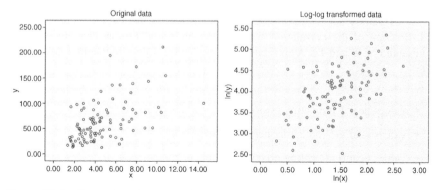

Figure 20.2   Examples of scatter plots of original data and logarithmically transformed data.

each state, the values along the horizontal axis are the abundances of protein one, and the values along the vertical axis the abundances of protein two.

- Is there a correlation between protein profiles at two specific sites, meaning, for example, that if a protein has high abundance at site one, it also has high abundance at site two? In this case there is one point for each protein.

Correlation analysis is mainly for studying the strength of correlation/relationships between two random variables. In this case one does not know if one of the variables may be the cause of the other. Also in the correlation analysis the X and Y variables are symmetrical, meaning that they can exchange axes.

The basic analysis considers *linear* correlation, the correlation is described by a straight line. The strength of the correlation is commonly measured by a *correlation coefficient*.

Generally, a correlation analysis is performed on a *sample* from a *population*, and the goal is to estimate the values of some *parameters* for the population by use of samples. Often the parameter is the correlation coefficient. To be able to use the correlation coefficients and the statistical analysis described here to make inferences about population correlation from sample correlation, the datasets should generally satisfy the following assumptions:

- The X and Y values are from populations that are approximately normally distributed, this is however less necessary for large samples.

- The samples are independently and randomly selected.

- The X and Y values are measured independently, and none is controlling the other. An example of controlled X-values is time, and an Y-value is measured for given time points. In this case regression should be used.

- The correlation holds true for the whole population.

As explained above a correlation coefficient is an indication of the strength and direction of the correlation between two independent variables (in our case represented by X and Y). A correlation coefficient has values in the interval $[-1, 1]$, with:

- values near 1 indicating positive correlation between the datasets;

- values near -1 indicating negative correlation;

- values near 0 indicating no correlation at all.

More detailed interpretation depends on the actual formula used for the correlation coefficient. We will briefly describe two of the most used coefficients.

20.8.1   Pearson's product-moment correlation coefficient

Pearson's correlation coefficient measures the degree of linear relationship between two random variables X and Y. In the population it is defined as

$$\rho(X, Y) = \frac{\sigma_{X,Y}}{\sigma_X \sigma_Y},$$

where $\sigma_{X,Y}$ is the covariance between X and Y. The covariance will be positive if Y increases with increasing X, and negative if Y decreases with increasing X. Its magnitude is, however, dependent on the scales of measurement for X and Y making it unsuitable for comparison of covariation when measurement units differ. ρ takes on values in the interval $[-1,1]$, negative values indicating that X increases with decreasing Y and positive values indicating increasing X with increasing Y. ρ will depend on the grade of linear relationship as well as on the variation of the joint distribution of X and Y, and is a unitless parameter. A perfect linear relationship will have $\rho = 1$ or $\rho = -1$, unless X or Y is constant.

Note that Pearson's correlation coefficient measures linear relationships, and there can be a perfect nonlinear relationship between X and Y with $\rho = 0$.

If we replace the population parameters with the corresponding sample parameters we get the *sample correlation coefficient*

$$R = \frac{1}{n-1} \sum_{i=1}^{n} \left(\frac{X_i - \bar{X}}{S_X} \right) \left(\frac{Y_i - \bar{Y}}{S_Y} \right),$$

where $X = (X_1, \ldots, X_n)$, $Y = (Y_1, \ldots, Y_n)$, and S_X and S_Y are the standard deviations of X and Y.

The Pearson correlation coefficient is invariant to changes in location and scale. This means that we can transform X to $aX + b$ and Y to $cY + d$, and the coefficient is the same.

When achieving a large value for R the reason for this should be further analyzed. There can be several reasons:

- the values of X are determined by the values of Y;

- the values of Y are determined by the values of X;

- there is another variable (Z) determining the values of both X and Y;

- the calculated value for R is high due to chance, there is a relation between X and Y in the sample, but not in the population.

To be able to say more about the last alternative we may test hypotheses about ρ.

Testing hypotheses about ρ

To make inferences about ρ we can formulate a null hypothesis:

$$H_0 : \rho = 0 \text{ versus } H_1 : \rho \neq 0.$$

For a chosen significance level α we can perform a two-sided test of H_0 in two different ways:

- Under H_0 the distribution for the test-statistic

$$T = R\sqrt{\frac{n-2}{1-R^2}}$$

  approximates a T distribution with $df = n - 2$.

- If $n \geq 10$

$$Z = 0.5 \ln \frac{1+R}{1-R} \sqrt{n-3}$$

  approximates a standard normal distribution (Motulsky (1995)).

Example For the example data in Figure 20.2, $R = 0.546$ and $n = 100$. Then $T = 3.18$, $Z = 2.86$. For both the T distribution and the standard normal distribution the p-values are less than 0.001, hence H_0 is rejected on the 0.001 level for both tests, and we conclude that there is a linear relation between the variables.
$\triangle$

Interpretation of R^2

R^2 has an important interpretation in that it is the proportion of variability in the Y-values that can be explained by the variability in the X-values. Given that the correlation coefficient is symmetric in X and Y, R^2 is also the proportion of the variability in X that can be explained by the variability in Y. R^2 is called *the coefficient of determination*.

Example For the example data in Figure 20.2, $R^2 = 0.298$ and almost 30% of the variability in the Y-values can be explained by the variability in the X-values.
$\triangle$

Confidence interval for ρ

Calculating a confidence interval for the *population* correlation coefficient ρ, from a *sample* correlation coefficient R, is far from straightforward.

Consider determining the confidence interval for ρ in a *normal* population, rewritten from Bhattacharyya and Johnson (1977). Let

$$L = \frac{1}{2} \ln \frac{1+R}{1-R}, \text{ and } \lambda = \frac{1}{2} \ln \frac{1+\rho}{1-\rho}.$$

For n sufficiently large we have:

$$P\left(L - \frac{z_{\frac{\alpha}{2}}}{\sqrt{n-3}} < \lambda < L + \frac{z_{\frac{\alpha}{2}}}{\sqrt{n-3}}\right) = 1 - \alpha.$$

A $100(1-\alpha)\%$ confidence interval for ρ can then be derived as:

$$\left(\tanh(L - \frac{z_{\frac{\alpha}{2}}}{\sqrt{n-3}}), \tanh(L + \frac{z_{\frac{\alpha}{2}}}{\sqrt{n-3}})\right),$$

where

$$\tanh(x) = \frac{e^{2x} - 1}{e^{2x} + 1}.$$

Example For the example with $n = 100$ and $R = 0.546$, we get $L = 0.613$. With $\alpha = 0.05$ we have $z_0.025 = 1.96$ and $z_0.05/rot(100 - 3) = 0.199$ and calculate the 95 % confidence interval to be

$$(\tanh(0.613 - 0.199), \tanh(0.613 + 0.199) = (\tanh(0.414), \tanh(0.812))$$
$$= (0.392, 0.671).$$

$\triangle$

Uncentered Pearson correlation coefficient

When we assume that the means of the populations are zero we get the uncentered Pearson correlation coefficient:

$$R = \frac{1}{n-1} \sum_{i=1}^{n} \frac{X_i Y_i}{S_X S_Y}.$$

20.8.2   Spearman's rank correlation coefficient

Spearman's rank correlation is nonparametric, meaning that no underlying distribution for the variables represented by the datasets is assumed. Spearman's correlation coefficient is a measure of the degree of monotonic relationship between two variables. Any perfectly monotonic relationship will give Spearman's correlation equal to 1 (or -1).

As the name implies the coefficient is based on ranks, see Section 20.4.2. It is easy to show that the average rank is

$$\bar{R} = \frac{n+1}{2}, \text{ and that } \sum_{i=1}^{n}(R_i - \bar{R})^2 = \frac{n(n^2-1)}{12}.$$

Spearman's correlation coefficient is defined analogously to Pearson's correlation coefficient by replacing the values by their ranks. Letting $(R_1, \ldots, R_n)$ be the ranks of the X data, and $(T_1, \ldots, T_n)$ the ranks of the Y data we get the Spearman's correlation coefficient R_S as:

$$R_S = \frac{1}{n-1}\sum_{i=1}^{n}\left(\frac{R_i - \bar{R}}{S_R}\right)\left(\frac{T_i - \bar{T}}{S_T}\right) = \frac{\sum_{i=1}^{n}(R_i - \bar{R})(T_i - \bar{T})}{\sqrt{\sum_{i=1}^{n}(R_i - \bar{R})^2}\sqrt{\sum_{i=1}^{n}(T_i - \bar{T})^2}},$$

which by use of the formulas above can be rewritten as:

$$R_S = \frac{\sum_{i=1}^{n}(R_i - \frac{n+1}{2})(T_i - \frac{n+1}{2})}{n(n^2-1)/12}.$$

We also derive

$$\sum_{i=1}^{n}\left(R_i - \frac{n+1}{2}\right)\left(T_i - \frac{n+1}{2}\right) = \sum_{i=1}^{n}R_iT_i - n\left(\frac{n+1}{2}\right)^2.$$

Further, using that $\sum_{i=1}^{n}i^2 = \frac{n}{6}(2n^2 + 3n + 1)$ we find

$$R_S = 1 - \frac{6}{n(n^2-1)}\sum_{i=1}^{n}(R_i - T_i)^2.$$

We see that for two datasets with identical rank values $R_S = 1$, and for datasets with completely opposite values $R_S = -1$.

Generally a Spearman rank correlation coefficient near 1 indicates that larger values of X are associated with larger values of Y, and a Spearman rank correlation coefficient value near -1 indicates that smaller values of X are associated with larger values of Y. This means that increasing/decreasing relationships can be discovered without the relationships being linear.

20.8.3   Correlation line

If there is a linear correlation between X and Y one is often interested in expressing this correlation as a straight line in the scatter plot that best illustrates the correlation. This is called the *correlation line*. One way of constructing such a line is to use *total least square*. If we plot the X, Y values in a diagram, it is the line that minimizes the

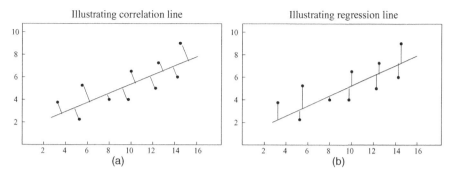

Figure 20.3   Illustration of how to calculate correlation line and linear regression. (a) Calculating a correlation line, where the orthogonal distances are used. (b) Linear regression, where the vertical distances are used. Note that the points are the same while the lines are slightly different.

sum of the squares of the orthogonal distances from the point to the line. This is also known as *orthogonal linear regression*, and is illustrated in Figure 20.3(a).

20.9   Regression analysis

Regression analysis is used to investigate if a random variable Y depends on another variable X. We here consider *linear* regression analysis, which means that we consider two variables:

- an *independent* variable, X, which sometimes is a random variable and sometimes has values that are controlled by the experimenter;

- a *dependent* variable, Y, for which the value depends on X.

The task is then to investigate if there is a linear dependency between X and Y through a formula

$$Y = \alpha + \beta X,$$

and to make estimates for α and β.

An example is if we want to examine if the abundance of a specific protein (Y) depends linearly on the amount of a specific drug (X). For a set of X-values we measure the Y-values after a specific time interval.

Regression analysis and correlation analysis are closely connected. The differences are mostly in terms of purpose and interpretation. While correlation does not distinguish between a predictor and a response variable, this is essential in regression analysis. Correlated variables will always be mutually dependent, but two dependent variables do not have to be correlated.

In regression analysis the purpose is to estimate or predict the values or expected values of the response variable from known values of the predictor variable. If we are postulating a causal relationship where one variable causes the other, a regression analysis is appropriate. However if there is an underlying, not observed, variable causing changes in both the two observed variables, a correlation analysis is better (or a so called *latent factor analysis*).

20.9.1   Regression line

The statistical model for examining linear dependency can now be formulated as:

- A relation $Y_i = \alpha + \beta X_i + U_i$, $i = 1, ..., n$ is assumed; where

  - $\{X_i\}$ is the set of values for the independent variable;

  - $\{Y_i\}$ is the set of respective measured values for the dependent variable;

  - $\{U_i\}$ is the set of unknown errors related to the true linear relation. These are assumed to be independent of each other, and normally distributed with mean zero and an unknown, but constant, variance;

  - α and β are the parameters to be estimated.

α is termed the *intercept* or the *constant*, and is the mean value for Y given $X = 0$. β is termed the *slope* or the *regression coefficient*, and is interpreted as the change in the mean of Y given one unit change in X. The commonly used method for estimating the true relation is the *least squares method*. Among all possible lines $Y = a + bX$ it finds values for a and b that minimize the sum of the squared vertical distances between the observed Y-values and the line given by the expression

$$\sum_{i=1}^{n} (Y_i - a - bX_i)^2.$$

Figure 20.3(b) illustrates the least squares method.

When the estimated values for a and b have been determined by the least squares method, one can calculate the difference between the measured values $\{Y_i\}$ and the values given by the line as $\hat{U}_i = Y_i - a - bX_i$. These $\{\hat{U}_i\}$ are called *residuals*.

Figure 20.4 shows both original data (same as Figure 20.2) and log-transformed data at both axes.

For the example data in Figure 20.4(a) $a = 3.034$ and $b = 0.664$. The interpretation here is that the mean value for Y increases with 0.664 units per one unit change in X, and that for $X = 0$ the mean value for Y is 3.034. Sometimes α has no real interpretation as 0 may be outside the plausible range of values for X and is only a nominal anchor point for the regression line. A problem with these data is that the variance of the error terms are not constant, but increases with increasing X-values. Also, both variables are highly right-skewed, and for the regression of Y on X the residuals are also right-skewed indicating not-normally distributed errors.

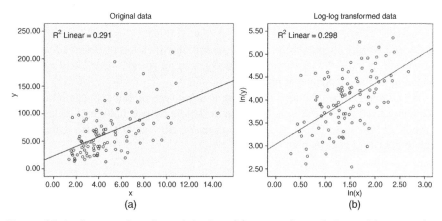

Figure 20.4   Scatter plots for original and log-transformed data with regression lines.

For the log-transformed data in Figure 20.4(b) the regression line is shown, where $a = 23.379$ and $b = 8.635$, $R = 0.546$ and $R^2 = 0.298$. The p-value for test of correlation is still < 0.001. Notice that this is equivalent to testing if the slope is 0. In this case $\beta \cdot 100$ is interpreted as the approximate percentwise increase in Y given $1\,\%$ increase in X. For the transformed data the variance of the error terms does not increase with increasing values for $\ln X$ and is thus more in line with the assumption for the model. Note that the two models based on Figures 20.4(a) and (b) are substantially different.

20.9.2   Relation between Pearson's correlation coefficient and the regression parameters

There are exact arithmetical relationships between the estimated Pearson's correlation coefficient R and the regression parameters given as $b = RS_Y/S_X$ and $a = \bar{Y} - b\bar{X}$.

We can also explain R^2 as being the fraction of the variance in Y that can be explained by a linear regression line with respect to X. This can be understood by rewriting R^2 as:

$$R^2 = 1 - \frac{S^2_{Y|X}}{S^2_Y},$$

where $S^2_{Y|X}$ is the square error of the linear regression of X on Y by $Y = a + bX$:

$$S^2_{Y|X} = \frac{1}{n-1} \sum_{i=1}^{n} (Y_i - a - bX_i)^2.$$

For the data in Figure 20.4(a) $R = 0.539$ and $R^2 = 0.291$, hence the regression line explains $29.1\,\%$ of the variance in Y.

20.10   Types of values and variables

At the end of this chapter we present a general classification of observed data and corresponding variables. A variable is usually used for storing the value of a feature, for example, the abundance of proteins.

A variable can be classified according to what type of values it can attain. One frequently used classification is based on an article by psychologist Stanley Smith Stevens, *On the theory of scales of measurement*, Stevens (1946). He defined four scale types of measurements.

Nominal scale   means that the possible values are discrete codes. An example is hair colour (auburn, black, blond, brown, grey, white, red). No ordering exists between the values, and it is meaningless to perform any quantitative comparison on such values. (If an ordering is defined for the hair colors, it becomes an ordinal scale variable.) The mode (the value that occurs most frequently) can be calculated, but not the mean or median. Data on the nominal scale are called *categorical* data, and variables of this scale are also called *categorical* variables.

Ordinal scale   means that an *ordering* among the values exists. An example is the grades for a test (A, B, C, D, E, F). It is, however, meaningless to perform arithmetic operations on them, for example, subtraction. This means that if one wants to present an 'average' result for a test, the 'average' should be the median, or the marks should be transformed to numbers before taking the mean. But then the value type is no longer ordinal.

Interval scale   means that it is meaningful to take the difference between values. An arbitrary zero point is defined, and negative values may exist. There is also a unit associated with the values. The most used example for interval scales is temperature measured in Celsius, where the zero point is the freezing point of water in atmospheric pressure, 100 degrees is the boiling temperature, thus the unit of measurement is defined. Note that it is not meaningful to take the ratio between two values, so division and multiplication are not performed. However, one can take ratios of temperature differences.

Ratio scale   means that it is also meaningful to take the ratio between values, examples of this are length and height. The zero point is not arbitrary.

It should be noted that there is some debate regarding this classification, and variants and extensions exist, see for example Velleman and Wilkinson (1993).

Using the classification of random variables used by statisticians, that is, into discrete and continuous variables, we may say that a nominal variable is a discrete variable with no ordering, an ordinal variable is a discrete variable with ordering but no scale, an interval variable is a variable with a scale but no fixed zero point, and a ratio scale variable is a variable with a scale and a fixed zero point. A count variable, for example, the number of children, is a discrete variable on a ratio scale. A mixed variable is a variable that can take on some values with positive probability and also values on a continuous scale, for example, the diameter of a tuberculin test can be 0 for a proportion of the population and continuous from $(0, \infty)$ for the rest of the population.

Which statistics that can be used on a variable depends on the type of variable. All statistics can be performed on variables of ratio scale. Means cannot be calculated for variables of nominal or ordinal scale, but medians can be found for ordinal variables. The mode can be determined for all types. Sometimes ordinal variables are analyzed as if they were continuous, and sometimes continuous variables are categorized and analyzed as ordinal or nominal variables.

21

Appendix B: Clustering and discriminant analysis

Clustering and discriminant analysis deals with objects (or events) and classes to which the objects belong. The field of clustering and discriminant analysis is large, and it is employed in various scientific fields such as statistics, machine learning, and pattern recognition. Each scientific field often uses different terms for the same entities, for example, other terms for class are group, cluster, category, and set. We will occasionally use these terms.

Both clustering and discriminant analysis starts with a set of objects, often called an *example set* or a *training set*. The difference between clustering and discriminant analysis can be described as:

- **Clustering:** The classes are unknown, and the aim is to organize the objects into classes, using the features of the objects.

- **Discriminant analysis:** The classes are known, and the class of each object in the training set is known. The objective is to achieve a description of each class, using the features.

In both cases one should afterwards be able to classify new objects into the correct classes. Discriminant analysis can also be used as a way of judging the result of clustering, thus the two complement each other.

21.1 Clustering

The main task in clustering is to divide the objects under consideration into reasonable classes, such that a structural ordering of the objects can be achieved.

Computational and Statistical Methods for Protein Quantification by Mass Spectrometry,
First Edition. Ingvar Eidhammer, Harald Barsnes, Geir Egil Eide and Lennart Martens.
© 2013 John Wiley & Sons, Ltd. Published 2013 by John Wiley & Sons, Ltd.

Example Consider a set of proteins, for which the abundances are measured in a time series. The objects are the proteins, and the features are the abundances measured at each time point. A protein's profile consists of all its measured abundances. The goal is to collect all proteins with 'similar' profiles in the same class, while proteins with 'different' profiles end up in separate classes. This can, for example, be of help in determining the functions of the proteins.

△

Generally one tries to make the classes such that the similarity of the objects inside a class is maximized, and the similarity between objects in different classes is minimized.

A clustering task consists of several subtasks:

- Which features should be used in the clustering?
- How should the proximity (distance or similarity) between the objects be measured?
- What should reasonable classes mean?
- How should the clustering be performed?

In addition one has to decide the following:

- Are the classes to be disjoint, or can an object belong to several classes?
- Is the number of classes fixed at startup, or can it be determined during the clustering process?

At the end the result also has to be validated.

In the following we consider clustering under the restrictions that the classes should be disjoint, there should be no empty classes, and all objects should belong to a class.

21.1.1   Distances and similarities

Distance and similarity are dualistic measures. A similarity measure is a numerical function of two objects, such that higher values mean larger similarities. A distance measure is an opposite numerical function, the larger the similarity the smaller the value. Roughly we can say that the distance measures the difference between the objects. Thus in many cases similarity and distance are interchangeable.

The form of a distance or similarity measure depends on the type or scale of the features, see Chapter 20.10. We here only consider interval scale features, and mainly use distances for the proximity. More about distance and similarity methods can, for example, be found in Theodoridis and Koutroumbas (2006).

21.1.2 Distance measures

A distance *measure d* for a set of objects X must satisfy:

1. d is a real function;

2. $d(x, x) = 0, \forall x \in X;$[1]

3. $d(x, y) = d(y, x), \forall x, y \in X.$

d is a *metric* if it in addition satisfies:

1. $d(x, y) > 0, \forall y \neq x \in X;$

2. $d(x, y) \leq d(x, z) + d(z, y), \forall x, y, z \in X$ (the triangular inequality).

L_p metric

The most used distance measure between objects with m interval scale features is the *weighted L_p* metric:

$$d_p(x, y) = \left(\sum_{i=1}^{m} w_i |x_i - y_i|^p \right)^{1/p},$$

where $x_i(y_i)$ is the i'th feature of $x(y)$, and $w_i \geq 0$ is a weight of feature i. In many cases $w_i = 1$ for all features, thus it is unweighted.

Different measures result from different values of p, the most common is the *Euclidean distance* where $p = 2$. The distance when $p = 1$ is called the *Manhattan distance*, and d_∞ is defined as $d_\infty(x, y) = \max_{1 \leq i \leq m}(w_i |x_i - y_i|)$.

Example If we have objects with two features $x = (4, 2)$, $y = (3, 5)$, and $w_i = 1$, then $d_1(x, y) = 4$, $d_2(x, y) = 3.2$, $d_\infty(x, y) = 3$.
$\triangle$

Examples for abundance profiles

Yona et al. (2006) describe three ways of measuring distances for protein profiles.

Normalized Euclidean metric[2]

$$d(x, y) = \sqrt{\frac{1}{m} \sum_{i=1}^{m} (x_i - y_i)^2}$$

---

[1] 0 can be exchanged by a minimum fixed value d_0.

[2] Note that this is different from normalized Euclidean distance.

Often some x_i or y_i are missing (see Section 21.1.6), such that the number of features taking part in the calculation can vary, therefore the values should be normalized by m.

Direction distance   This is used to compare the change in abundances in two proteins, for example in a time series. Define

$$\hat{x}_i = \begin{cases} +1 & \text{if } x_{i+1} < x_i \\ -1 & \text{if } x_{i+1} > x_i \\ 0 & \text{if } x_{i+1} = x_i \end{cases}$$

The distance is then calculated as

$$d(x, y) = \frac{1}{2m} \sum_{i=1}^{m} |\hat{x}_i - \hat{y}_i|$$

Derivative distance   Similar to the direction distance, but $\hat{x}_i(\hat{y}_i)$ is defined as $\hat{x}_i = x_{i+1} - x_i (\hat{y}_i = y_{i+1} - y_i)$.

21.1.3   Similarity measures

The most common similarity measures for interval scale features are the Pearson correlation-coefficient and Spearman rank correlation coefficient (see Chapter 20.8), and the normalized *inner product* $s(x, y) = \frac{1}{m} \sum_{i=1}^{m} x_i y_i$.

21.1.4   Distances between an object and a class

During the clustering process one often has to decide to which class C an object x should be assigned. This will be decided using the distance (or similarity) between x and C. One must therefore be able to measure such a distance. Two approaches are mainly used:

1. Calculate and use the distances between x and each object y in C. Then for the distance between x and C use:

   - the maximum distance

$$d_{max}(x, C) = \max_{y \in C} d(x, y);$$

   - the minimum distance

$$d_{min}(x, C) = \min_{y \in C} d(x, y);$$

- the average distance

$$d_{ave}(x, C) = \frac{1}{|C|} \sum_{y \in C} d(x, y).$$

2. Calculate a representative (virtual) object for C, and measure the distance from x to it. Typically the representative object is the mean or median object, but can also be a geometrical center. Note that the representative object has to be recalculated each time a class changes.

Note that the definition of measure and metric is not reasonable to use for distances between an object and a class, since the two values are of different types (object and class).

21.1.5   Distances between two classes

During a clustering process one often has to decide if two classes should be joined or split. A way of measuring the distance (or similarity) between two classes is therefore needed. Again we consider the distances between the objects of the classes, and we have several alternative measures.

- The maximum distance:

$$d_{max}(C_i, C_j) = \max_{x \in C_i, y \in C_j} d(x, y)$$

This is not a measure, since $d_{max}(C_i, C_i)$ is not necessarily equal to 0.

- The minimum distance:

$$d_{min}(C_i, C_j) = \min_{x \in C_i, y \in C_j} d(x, y)$$

This is a measure, but not a metric since the triangle inequality is not necessarily satisfied.

- The average distance:

$$d_{ave}(C_i, C_j) = \frac{1}{|C_i||C_j|} \sum_{x \in C_i} \sum_{y \in C_j} d(x, y)$$

This is not a measure since $d_{ave}(C_i, C_i)$ is not necessarily equal to 0.

- A representative object:

$$d_{repr}(C_i, C_j) = d(r_{C_i}, r_{C_j}),$$

where r_{C_i} is the representative for C_i. This is a measure if d is a measure. But it is not a metric, since it can be zero for $C_i \neq C_j$.

For a specific clustering algorithm the resulting classes depend on the distance (or similarity) function used.

21.1.6 Missing data

Often it is impossible or not appropriate to measure the values of all features for all objects, for example, the abundance of all proteins. In such cases it is not possible to measure the distance between two objects using all features. Several ways of handling missing data have been proposed, the most common are:

1. ignore the features that contain missing data;

2. replace a missing value with the average of all measured values for that feature;

3. impute the missing value based on additional information, for example, the average or median from similar profiles.

Example We measure the abundances of four proteins in five situations $P_1 = (10, 7, 12, 6, 0)$, $P_2 = (7, ?, 9, 3, 2)$, $P_3 = (9, 5, 7, ?, 5)$, $P_4 = (6, 3, 4, 7, 2)$, where ? represents missing data. We can then either ignore features two and four in all calculations (needs no normalization), or only ignore them when p_2 or p_3 are included (needs normalization), or, for example, replace ? in P_2 by the average of the other proteins in situation 2, 5 (needs no normalization). Normalization is here considered necessary if the number of values taking part in the distance calculations can be different.

$\triangle$

21.1.7 Clustering approaches

In order to get an idea of the complexity of the clustering problem look at how many different ways a set of n objects can be clustered into k classes:

$$\frac{1}{k!} \sum_{i=0}^{k} (-1)^{k-i} \binom{k}{i} i^n$$

For $n = 9$, $k = 4$ there are 7770 different clusterings. When k is not fixed, but allowed to be a value in $[1, n]$, we see that the number becomes large even for small n.

There are a lot of different clustering approaches, of which the most common are:

- sequential clustering;

- hierarchical clustering;

- k-means clustering.

We here briefly describe the first two. In these approaches the number of clusters are not *a priori* given. For k-means the number of clusters (k) is given by the user.

In the algorithms we use d for distances, also between an object and a class, and between two classes. We also use q instead of x for an object. The reason for this is that we will also use q_i for an object, and x_i is often used to specify features.

21.1.8   Sequential clustering

Sequential clustering works by first creating a class containing one of the objects, and the rest of the objects are then considered sequentially. For each object one has to decide if it should be assigned to an already created class, or if a new class should be created for the object. Typically the user provides two input values:

- the maximum number of classes, h;

- the distance threshold ϕ for deciding if an object q should start a new class or not.

If the distance from q to all of the created classes is larger than ϕ, a new class consisting of q is created (as long as the number of classes is less than h).

The basic algorithm is quite simple:

Algorithm 21.1.1   Sequential clustering algorithm

arguments
h *max. number of classes*
ϕ *distance threshold*
$Q = \{q_1, \dots, q_n\}$ *the objects*
var
m *the number of classes*
begin
   $m := 1; C_p := \{q_1\}$
   for $i = 2$ **to** n **do**
      $k := \arg \min_{1 \le j \le m} d(q_i, C_j)$ *the nearest class to q_i*
      if $d(q_i, C_k) > \phi$ **and** $m < h$ **then**
         $m := m + 1; C_m := \{q_i\}$ *new class*
      else
         $C_k := C_k \cup \{q_i\}$
         update possible representative for C_k
      end
   end
end

Example   Assume that we have six objects with two features:

$$q_1 : (2, 2); \quad q_2 : (2, 4); \quad q_3 : (4, 8); \quad q_4 : (8, 5); \quad q_5 : (6, 9); \quad q_6 : (5, 3).$$

We use Euclidean distance, and find the distances between each object as:

	q_2	q_3	q_4	q_5	q_6
q_1	2.0	6.3	6.7	8.1	3.2
q_2		4.5	6.1	6.4	3.2
q_3			5.0	2.2	5.1
q_4				4.5	3.6
q_5					6.1

Let us use $d_{ave}(q, C)$ as the distance between an object and a class, and use $\phi = 4$. If the objects are presented in the given order, the algorithm will produce the following:

$C_1 = \{q_1\}$;
$C_1 = \{q_1, q_2\}$;
$C_2 = \{q_3\}$ since $d(q_3, C_1) = 5.4$;
$C_3 = \{q_4\}$ since $d(q_4, C_1) = 6.4$, $d(q_4, C_2) = 5$;
$C_2 = \{q_3, q_5\}$ since $d(q_5, C_2) = 2.2$;
$C_1 = \{q_1, q_2, q_6\}$ since $d(q_6, C_1) = 3$;

Thus the final clustering is

$$C = \{\{q_1, q_2, q_6\}\{q_3, q_5\}\{q_4\}\}.$$

If the objects are presented in the order $q_3, q_4, q_6, q_1, q_2, q_5$ the clustering will be

$$C = \{\{q_3, q_5\}\{q_4, q_6\}\{q_1, q_2\}\}.$$

△

A couple of remarks regarding the above algorithm:

- Only one clustering is found and displayed to the user.

- The clustering result depends on the distance measure used.

- The values of ϕ and h affect the clustering.

- The clustering depends on the order in which the objects are presented.

- The class to which an object is attached only depends on the already clustered objects.

Variations of this basic algorithm have therefore been proposed. Below are three options related to the two last points.

- A two step procedure, where in the first step one performs preliminary class assignments of some of the objects, each class consisting of exactly one object. These consist typically of objects being far from each other. The assignment of all the other objects are done in the second step.

- An iterative procedure, where in each iteration cycle one only assigns those objects where one asserts with high certainty which class the objects should be assigned to. This is typically performed by using two threshold arguments, ϕ_1 and ϕ_2. In each iteration cycle an object that is not already assigned to a class will be treated as one of three alternatives:

- assigned to a class if the distance to the class is $< \phi_1$;

- a new class is created if the distances to all existing classes are $> \phi_2$;

- the object is still unassigned.

- A refinement step can be included after the clustering to join classes that are close to each other, and move objects to other classes where they seem to fit better.

21.1.9   Hierarchical clustering

These algorithms produce a set of clusterings, instead of just one, and can roughly be divided into two types:

- *Agglomerative algorithms*   One starts with n (disjoint) classes, each consisting of one object. This is the clustering on level 1. An iterative procedure is then performed, where two classes of the current clustering are joined in each iteration cycle. At the end one has a clustering consisting of one class with all the objects.

- *Divisive algorithms*   One starts with one class consisting of all objects, and in each iteration cycle one of the current classes is split into two classes.

Note that all the created clusterings are available, thus a hierarchy of different clusterings can be considered.

Algorithm 21.1.2 shows a generic algorithm for the agglomerative approach.

Algorithm 21.1.2   Agglomerative clustering

arguments
$Q = \{q_1, \ldots, q_n\}$ *the objects*
var
C_i *a clustering on level i*
begin
    $\mathcal{C}_1 := \emptyset$
    for $i = 1$ **to** n **do**   *one class of each object*
        $C_i := \{q_i\}; \mathcal{C}_1 := \mathcal{C}_1 \cup \{C_i\}$
    end
    for $i = 2$ **to** n **do**   *make a new clustering in each iteration cycle*
                                by joining two classes
        $r, s := \arg \min_{C_j, C_k \in \mathcal{C}_{i-1}} d(C_j, C_k)$
        $C_q = C_r \cup C_s$
        $\mathcal{C}_i := \mathcal{C}_{i-1} - \{C_r, C_s\} \cup \{C_q\}$
    end
end

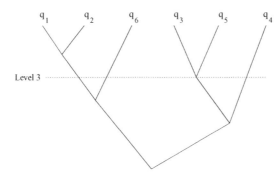

Figure 21.1   Tree illustrating the example clusterings. At level one there are 6 classes, at level three there are 4.

Depending on how the different steps are implemented one gets different clusterings. Figure 21.1 shows an example of hierarchical clustering.

The class distances

In each iteration cycle of the algorithm we need the distances between the classes in clustering C_i. When two classes C_r, C_s are joined into a new cluster C_q we need to calculate the distances to C_q from all the other classes C_o. This can be performed in a number of ways, but will generally depend on the distances $d(C_q, C_r), d(C_q, C_s), d(C_r, C_s)$, for example as a linear function.

We here present four frequently used calculations where $d(C_r, C_s)$ is not taken into account.

Weighted Pair Group Method Average – WPGMA

$$d(C_q, C_o) = \frac{1}{2}(d(C_r, C_o) + d(C_s, C_o))$$

Unweighted Pair Group Method Average – UPGMA

$$d(C_q, C_o) = \frac{1}{m_r + m_s}(m_r d(C_r, C_o) + m_s d(C_s, C_o)),$$

where m_r is the number of objects in C_r. This corresponds to $d_{ave}(C_i, C_j)$ in Section 21.1.5.

Single link

$$d(C_q, C_o) = min(d(C_r, C_o), d(C_s, C_o))$$

This corresponds to $d_{min}(C_i, C_j)$ in Section 21.1.5.

Complete link

$$d(C_q, C_o) = max(d(C_r, C_o), d(C_s, C_o))$$

This corresponds to $d_{max}(C_i, C_j)$ in Section 21.1.5.

The single link method has a tendency of forming elongated classes, while the complete link method has a tendency for compact classes.

Example Consider the same set of objects as in the example in Section 21.1.8. For the distances we use complete link (d_{max}). In each iteration cycle the classes with the smallest distance are joined.

	q_2	q_3	q_4	q_5	q_6
q_1	2.0	6.3	6.7	8.1	3.2
q_2	0	4.5	6.1	6.4	3.2
q_3		0	5.0	2.2	5.1
q_4			0	4.5	3.6
q_5				0	6.1

$\Rightarrow$

	q_3	q_4	q_5	q_6
$\{q_1, q_2\}$	6.3	6.1	8.1	3.2
q_3		5.0	2.2	5.1
q_4			4.5	3.6
q_5				6.1

$\Rightarrow$

	$\{q_3, q_5\}$	q_4	q_6
$\{q_1, q_2\}$	8.1	6.1	3.2
$\{q_3, q_5\}$		5.0	6.1
q_4			3.6

$\Rightarrow$

	$\{q_3, q_5\}$	q_4
$\{q_1, q_2, q_6\}$	8.1	6.1
$\{q_3, q_5\}$		5.0

$\Rightarrow$

	$\{q_3, q_5, q_4\}$
$\{q_1, q_2, q_6\}$	8.1

The clusterings are shown in Figure 21.1.
The set of clusterings at each level are:

$$\mathcal{C}_2 = \{\{q_1, q_2\}\{q_3\}\{q_4\}\{q_5\}\{q_6\}\} \quad \mathcal{C}_3 = \{\{q_1, q_2\}\{q_3, q_5\}\{q_4\}\{q_6\}\}$$

$$\mathcal{C}_4 = \{\{q_1, q_2, q_6\}\{q_3, q_5\}\{q_4\}\} \quad \mathcal{C}_5 = \{\{q_1, q_2, q_6\}\{q_3, q_5, q_4\}\}$$

$$\mathcal{C}_6 = \{\{q_1, q_2, q_6, q_3, q_5, q_5\}\}$$

We see that one of the clusterings in the example using the sequential algorithm is the same as clustering $\mathcal{C}_4$.
$\triangle$

Best number of classes

When a hierarchy of clusterings is obtained, one wants to present the most 'correct' clustering, that is, when should we stop the clustering process? One way of determining this is to inspect the properties of the classes, for example, consider the *lifetime* of a class: The number of iteration cycles between its creation and when it is joined with another class. Classes with long lifetimes are considered to be 'good.' Other more advanced methods can be found in Theodoridis and Koutroumbas (2006), where other aspects related to clustering are also discussed, for example, validation.

21.2   Discriminant analysis

Discriminant analysis considers objects where it is known which class each object belongs to. Briefly we can say that discriminant analysis is the study of differences between classes by considering their features.[3]

The task we will cover here can be described as:

- Determine the features that best discriminate between the classes, *feature selection* or *feature extraction.*

- Determine how these features can be combined to best determine (predict) to which class an unknown object belongs.

These two subtasks can be performed separately or in a combined procedure.

Example As an example from protein quantification let the objects be humans, and the features be the abundances of a set of proteins. The classes will be the situations, for example, healthy individuals, individuals in the early stages of a disease, and individuals in advanced stages of the same disease. The goal is to find a set of proteins for which the abundances can be used to discriminate between the humans from the three situations, that is, finding a panel of biomarkers for the disease. △

There are many different approaches and methods for discriminant analysis, and the literature is mostly related to statistics and machine learning. Some methods from the field of machine learning are decision tree learning, concept learning algorithms, artificial neural networks, support vector machines, and Bayesian learning. In machine learning the known objects are called the *training set* (or training examples), and it is important that they are representative for the class they belong to, and that they are suitable for learning the distinction between the classes. The statistical methods are often closely related to regression and correlation analysis, ANOVA/MANOVA, principal component analysis, and factor analysis.

The various approaches often use different terms and concepts, and the same term may have different meanings in the literature or statistical packages. We try to use the most common terms and meanings, however a given term can have other meanings in some literature/packages, or may not be used at all.

---

[3] The features are often termed variables or discriminating variables in the context of discriminant analysis.

21.2.1 Step-wise feature selection

Step-wise feature selection is a simple way of identifying the features that seems to be best suited for describing the differences between the classes. We can consider two variants of this task:

1. Given a set of m features, find the subset that 'best describes the discrimination.'

2. Given a set of m features, find the 'best' subset of size d.

This problem is often termed feature selection. To solve it we need a formula for measuring the *discrimination power* of a set of features, and then measure each possible subset. This can however be cumbersome, the number of different subsets of m features is 2^m, and the number of subsets of size d is

$$\frac{m!}{(m-d)!d!}.$$

There are numerous measures for discrimination power, see for example Klecka (1980). A common option is using *Wilks' Lambda*. It takes into account the differences between the classes, and the similarity (homogeneity) within the classes. For testing the significance Wilks' Lambda is often transformed to an F-distributed statistic.

Different search algorithms have been developed, most of them being heuristic, that is, not guaranteed to find the optimal solution. In the following we consider three different step-wise algorithms.

Forward   One starts with zero features, and in a step-wise manner adds new features, picking the feature with highest discriminatory power together with the already selected features. At each step one investigates if adding the feature substantially increases the discriminatory power of the subset.

Backward   One starts with all features, and in a step-wise manner removes features that, when removed, do not decrease the discriminatory power substantially.

Combined Forward/Backward   (Often called *F to enter, F to remove.*) A drawback with the above approaches is that when one feature is added or removed, this cannot be restored. The combined Forward/Backward procedure allows this, which means that a more 'overall view' is used when determining the best subset.

Below we present rough algorithms for the three approaches. In the algorithms we use two procedures:

- *signif_when_add(x, X)*, returning a calculated significance of the increase in discriminatory power when feature x is added to the set X. If the discriminatory power is decreased it returns zero. Note that the significance here is a positive value, the highest value showing highest significance.

- *signif_when_remove(x, X)*, returning a calculated significance of the decrease in discriminatory power when feature x is removed from X. If the discriminatory power is increased it returns zero.

The algorithms for Forward and Backward are straightforward. The Combined Forward/Backward approach can be implemented in different ways. We have chosen to have a main iteration for the forward approach, and start each iteration cycle by testing if one (and only one) of the features in the running subset should be removed.

Algorithm 21.2.1   Forward step-wise discriminant analysis

proc
signif_when_add(x, X) returns the significance of the increase
*   in discrimination power when feature x is added to the set X*
const
dp_limit_1   an x_i with significance increase above dp_limit_1 will be added to X
var
X   the set of original features
Z   will store the 'best' subset of original features
dp_i significance of increase in discriminatory power
begin
   $X :=$ *the set of features*
   $Z := \emptyset$
   cont := true
   while *cont* **do**
     cont := false
     for each $x_i \in X$ **do** $dp_i = signif\_when\_add(x_i, Z)$ **end**
     $j := argmax_i(dp_i)$ *i over the elements in X*
     if $dp_j > dp\_limit\_1$ **then**
        $Z := Z \cup \{x_j\}; X := X - \{x_j\}$
        cont := true
     end
   end
   return(Z)
end

Algorithm 21.2.2   Backward step-wise discriminant analysis

proc
signif_when_remove(x, X) returns the significance of the decrease
*   in discriminatory power when feature x is removed from X*
const
dp_limit_2 x_i with significance decrease below dp_limit_2 will be removed from X

var
X   *start with the set of original features*
                  will get the new subset
dp_i *significance of decrease in discriminatory power*
begin
   X := the set of features
   cont := true
   while *cont* **do**
     cont := false
     for each $x_i \in X$ **do** $dp_i = signif\_when\_remove(x_i, X)$ **end**
     $j := argmin_i(dp_i)$
     if $dp_j < dp\_limit\_2$ **then**
       $X := X - \{x_j\}$
       cont := true
     end
   end
   return(X)
end

Algorithm 21.2.3   Combined Forward/Backward step-wise discriminant analysis

proc
signif_when_add(x, X)
signif_when_remove(x, X)
const
dp_limit_1
dp_limit_2
var
X   the set of original features
Z   will store the new subset of original features
dp_i *significance of increase/decrease in discrimination power*
begin
   X := the set of features
   $Z := \emptyset$
   cont := true
   while *cont* **do**
     cont := false
     for each $z_i \in Z$ **do** $dp_i = signif\_when\_remove(z_i, Z)$ **end**
     $j := argmin_i(dp_i)$
     if $dp_j < dp\_limit\_2$ **then**
       $Z := Z - \{z_j\}$
       cont := true
     end
     for each $x_i \in X$ **to** $dp_i = signif\_when\_add(x_i, Z)$ **end**
     $j := argmax_i(dp_i)$

if $dp_j > dp\_limit\_1$ **then**
    $Z := Z \cup \{x_j\}; X := X - \{z_j\}$
    $cont := true$
  end
  end
  return(Z)
end

Example Bohne-Kjersem et al. (2010) describe a search for biomarkers in Atlantic cod. Five different situations (classes in the terminology of discrimination) are considered, described by different levels of produced water (H,L,M), oestradiol (E2), and a control (C). A total of 19 individual cod (the objects) were examined for differences in protein abundances. Thus the features are the protein abundances. By use of 2D gel electrophoresis and mass spectrometry 84 proteins where found to be differentially abundant. By a step-wise discriminant analysis they were able to find three proteins for which the abundances could be used to correctly classify the 19 cods. This is shown in Figure 21.4(A). The figure is further explained in Section 21.2.3.
△

A challenge when using step-wise discriminant analysis is to define the limits for when to stop the iteration.

21.2.2   Linear discriminant analysis using original features

When original features are used (possibly after feature selection) one tries to determine functions of the features that can be used to distinguish between the classes. When the functions are linear combinations of the features this is called *linear discriminant analysis – LDA*.

Note that an object can be represented as a data point in an m dimensional space, where one feature corresponds to one dimension. This means that for $m = 1$ the separations are points, for $m = 2$ they are lines, for $m = 3$ they are planes, and for $m > 3$ they are hyperplanes.

Example Figure 21.2 illustrates a case for $m = 3$. Each object is described by three features determining the three coordinate values in the 3D space. Three different classes are illustrated by different geometrical figures. (a) shows that it is here possible to draw two parallel planes that separate the classes. Since they are parallel it is possible to make a projection line perpendicular to the two planes. When projecting the figures and the planes we see that the points where the planes are projected (marked by filled dots) separate the projected figures belonging to the different classes. (b) illustrates the projection by first projecting the figures on a plane perpendicular to the separation planes, and then on the projection line.
△

A formal description of the problem is that we have:

- a set of known classes;

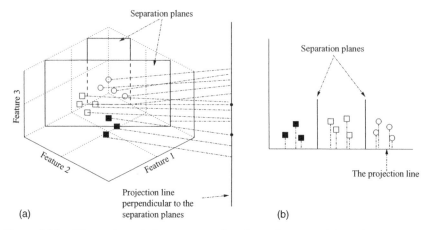

Figure 21.2   Illustration of planes separating three classes in 3D, and projection to a line.

- a set of known objects, and which classes each object belongs to;

- each known object is described by the values of a set of m features of interval scale, $x = \{x_1, \ldots, x_m\}$, where x can be considered as a vector.

The task is to derive a function of how the classes are separated, such that new objects can be reliably classified.

It is however important to realize that in many cases it is not possible to achieve a separation function with the method used that completely separates the classes.

The most common approach is to find a way of *minimizing the number of misclassified objects*. This is typically done by minimizing the *total error of classification, TEC*. TEC can roughly be described as the probability that the classification procedure will wrongly classify a random object.

A procedure for classifying a given object x, is for each class to calculate the probability that the object belongs to that class. The class with highest probability is then chosen.

The mathematical derivation of such a probability function is outside the scope for this book, but it is based on assuming that the data are multivariate normally distributed for each class, which means that every linear combination $y = a_1 x_1 + \cdots + a_m x_m$ is normally distributed. Note that this means that each feature x_i is normally distributed.

By use of Bayes' theorem and restricting to two classes we can derive the condition for x being classified to class 1 as

$$w \cdot x > k,$$

where

- w and x are considered as vectors;

- w and k is determined from the covariance matrices for the two classes, the mean value vectors for each class, and the prior probabilities for x being classified to class 1 or 2.

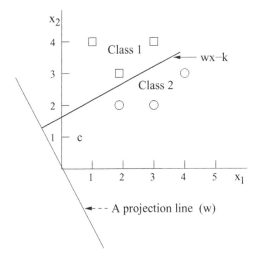

Figure 21.3   Illustration of a line separating two classes.

Geometric interpretation

Remember that the objects can be considered as data points in an m dimensional space, where each x_i represents one axis (dimension). This means that $w \cdot x = k$ is the equation for a hyperplane separating the two classes if they are linearly separable. w is perpendicular to the hyperplane, thus objects can be projected onto w, on which the classes are separated by a point. This is illustrated in Figure 21.3.

21.2.3   Canonical discriminant analysis

Above we have shown that in some cases a projection line can be found to discriminate (or separate) between classes. This means that we have transformed the objects from being represented by m features in an m dimensional space to be represented by one new feature in a one dimensional space. When using LDA the transformation is linear.

In most cases it is not possible to find one projection line to discriminate between the classes. We can however try to transform the original features to several new features, referred to as *feature extraction*. We define each new feature as a linear combination of the original features,

$$y_j = w_{j1}x_1 + \cdots + w_{jm}x_m,$$

and the number of such features defines the new dimension. The features $\{y_j\}$ are called *canonical features*, or *canonical components*. The coefficients $\{w_{ji}\}$ of the linear combination are called *canonical coefficients* or *canonical weights*. The canonical features are also called *canonical discriminant functions*, and the coefficients *canonical discriminant function coefficients*. We will use feature and function interchangeably, depending on what is most appropriate in the considering context.

The canonical discriminant analysis briefly described below is based on the following assumptions of the populations from which the data is collected:

- None of the original features can be written as a linear combination of any of the others.

- The covariance matrices are equal for each population class.

- The data comes from multivariate normal distributions for each class.

The canonical features are determined such that maximum separation of the classes can be obtained (and illustrated) by transforming the values of the original features to values of the canonical features.

Determining the canonical functions (features) is done by considering the variation within the groups, *within-group variation*, and the variation between the groups, *between-group variation*. One then tries to maximize the ratio r of the between-group variation to the within-group variation. The canonical functions are determined such that each is linearly independent of the others, that is, each function is orthogonal to each of the others, thus forming a coordinate system in the k dimensional space where k is the number of canonical functions.

The mathematical operation for finding the canonical functions is by forming linear equations, and solving for eigenvalues and eigenvectors. Each (eigenvalue, eigenvector) pair represent a canonical function y_j where the eigenvector $\{w_{ji}\}$ defines the coefficients, and the value of the eigenvalue λ_j says something about the discriminatory power of the function, the highest value means highest power. Thus the solution $(\lambda_1, \{w_{1i}\})$ with the highest eigenvalue called the *first* canonical feature, and so on.

There is a maximum of m (the number of original features) solutions to the linear equations mentioned above, but if m is greater than the number of classes $s - 1$ then only $s - 1$ of them has eigenvalues different from zero. This implies that the maximum number of canonical features is the minimum of $(m, s - 1)$. For example if there are seven original features, and four classes, the maximum number of canonical features is three.

Note that for m original features and k canonical features, there are $m \cdot k$ canonical coefficients. For each object these are then used to calculate the values of the canonical features, often called *scores*. The objects can be represented in a k dimensional coordinate system, with the scores of the objects being the values on the axes. For $k = 2$ this can be illustrated in a *canonical plot*, as shown in the following example.

Example In the article Bohne-Kjersem et al. (2010), mentioned in the example in Section 21.2.1, canonical discriminant analysis is used, illustrated in Figure 21.4.

In (B) there are 14 original features (protein abundances), and it is clear that the canonical plot of the first and second canonical features separates the five classes well. The score scales are shown on the axes. The canonical coefficients of the first canonical feature are also shown. In (A) there are only three original features, showing

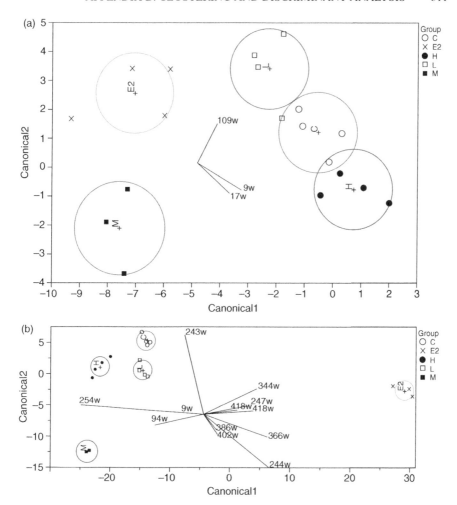

Figure 21.4   Examples of canonical plots. Reprinted from Bohne-Kjersem et al. (2010) by permission of Elsevier.

that using only three proteins separates the classes well by canonical discriminant analysis.

△

The approach for finding canonical coefficients as explained above is called *raw canonical coefficients*, and it is sufficient to illustrate discrimination. In more in-depth discriminant analysis one is interested in investigating the relation between the original and canonical features, quantitative values for the feature's impact on the differences between classes, the differences and correlations between classes, and so on. It is then appropriate to transform the coefficients such that the values of the *canonical features* are in a *standard* format. This means that the overall means of the scores for the objects are zero, such that the origin of the coordinate system contains the centroid of the set of objects. Also, the within-class standard deviation is one.

In the beginning of this section three requirements for the analysis were listed. Several authors have shown that the analysis is fairly robust, such that some deviation from the requirements can be tolerated. It is however important to always have these requirements in mind when interpreting the results.

21.3   Bibliographic notes

Clustering   This is described in a number of books, for example in Theodoridis and Koutroumbas (2006).

Discriminant analysis   This is also thoroughly described in various text books, and introductions can be found in, for example, Klecka (1980) and Thompson (1984).

Bibliography

Alberts B, Johnson A, Lewis J, Raff M, Roberts K and Walter P 2002 *Molecular biology of the THE CELL*. Garland Science.

Albrecht D, Kniemeyer O, Brakhage AA and Guthke R 2010 Missing values in gel-based proteomics. *Proteomics* **10**, 1202–1211.

Allet N *et al.* 2004 In vitro and in silico processes to identify differentially expressed proteins. *Proteomics* **4**, 2333–2351.

Alves P, Arnold RJ, Novotny MV, Radivojac P, Reilly JP and Tang H 2007 Advancement in protein inference from shotgun proteomics using peptide detectability *Pac. Symp. Biocomput.*, pp. 409–420.

Armbruster DA and Pry T 2008 Limit of blank, limit of detection and limit of quantitation. *Clin. Biochem. Rev.* **29**, S49–S52.

Armenta JM, Hoeschele I and Lazar IM 2009 Differential protein expression analysis using stable isotope labeling and pqd linear ion trap ms technology. *J. Am. Soc. Mass Spectrom.* **20**, 1287–1302.

Arntzen MØ, Koehler CJ, Barsnes H, Berven FS, Treumann A and Thiede B 2011 Isobariq: Software for isobaric quantitative proteomics using iptl, itraq, and tmt. *J. Proteome Res.* **10**, 913–920.

Aye TT, Low TY, Bjørlykke Y, Barsnes H, Heck AJ and Berven FS 2012 Use of stable isotope dimethyl labeling coupled to selected reaction monitoring to enhance throughput by multiplexing relative quantitation of targeted proteins. *Anal. Chem.* Published online: www.ncbi.nlm.nih.gov/pubmed/22548487.

Balasubramaniam D, Eissler CL, Stauffacher CV and Hall MC 2010 Use of selected reaction monitoring data for label-free quantification of protein modification stoichiometry. *Proteomics* **10**, 4301–4305.

Bantscheff M, Eberhard D, Abraham Y, Bastuck S, Boesche M, Hobson S, Mathieson T, Perrin J, Raida M, Rau C, Reader V, Sweetman G, Bauer A, Bouwmeester T, Hopf C, Kruse U, Neubauer G, Ramsden N, Rick J, Kuster B and Drewes G 2007a Quantitative chemical proteomics reveals mechanisms of action of clinical abl kinase inhibitors. *Nat. Biotechnol* **25**, 1035–1044.

Bantscheff M, Schirle M, Sweetman G, Rick J and Kuster B 2007b Quantitative mass spectrometry in proteomics: a critical review. *Anal. Bioanal. Chem.* **389**, 1017–1031.

Barsnes H, Vizcano JA, Eidhammer I and Martens L 2009 Pride converter: making proteomics data-sharing easy. *Nat. Biotechnol.* **27**, 598–599.

Beck HC, Nielsen EC, Matthiesen R, Jensen LH, Sehested M, Finn P, Grauslund M, Hansen AM and Jensen ON 2006 Quantitative proteomic analysis of post-translational modifications of human histones. *Mol. Cell. Proteomics* **5**, 1314–1325.

Benjamini Y and Hochberg Y 1995 Controlling the false discovery rate: a practical and powerful approach to multiple testing. *J. Royal Stat. Soc. B* **57**, 289–300.

Benjamini Y and Yekutieli D 2001 The control of the false discovery rate in multiple testing under dependency. *The Annals of Statistics* **29**, 1165–1188.

Bern M, Goldberg D, McDonald WH and Yates III, JR 2004 Automatic quality assessment of peptide tandem mass spectra. *Bioinformatics* **20(Suppl. 1)**, i49–i54.

Beynon RJ, Doherty MK, Pratt JM and Gaskell SJ 2005 Multiplexed absolute quantification in proteomics using artificial qcat proteins of concatenated signature peptides. *Nat. Methods* **2**, 587–589.

Bhattacharyya GK and Johnson RA 1977 *Statistical Concepts and Methods*. John Wiley & Sons, Ltd.

Bianco L, Mead JA and Bessant C 2009 Comparison of novel decoy database designs for optimizing protein identification searches using abrf sprg2006 standard ms/ms data sets. *J. Proteome Res.* **8**, 1782–1791.

Boehm AM, Pütz S, Altenhöfer D, Sickmann A and Falk M 2007 Precise protein quantification based on peptide quantification using itraq. *BMC Bioinformatics* **8**, 214–231.

Boersema PJ, Aye TT, van Veen TAB, Heck AJR and Mohammed S 2008 Triplex protein quantification based on stable isotope labeling by peptide dimethylation applied to cell and tissue lysates. *Proteomics* **8**, 4624–4632.

Bohne-Kjersem A, Bache N, Meier S, Nyhammer G, Roepstorff P, Sæle Ø, Goksøyr A and Grøsvik BE 2010 Biomarker candidate discovery in atlantic cod (gadus morhua) continuously exposed to north sea produced water from egg to fry. *Aquat. Toxicol.* **96**, 280–289.

Bolstad BM, Irizarry RA, Åstrand M and Speed TP 2003 A comparison of normalization methods for high density oligonucleotide array data based on variance and bias. *Bioinformatics* **19**, 185–193.

Booth JG, Eilertson KE, Olinares PDB and Yu H 2011 A bayesian mixture model for comparative spectral count data in shotgun proteomics. *Mol. Cell. Proteomics* **10**, 10.1074/mcp.M110.007203–1.

Box GEP and Cox DR 1964 An analysis of transformation. *J. Royal Stat. Soc. B* **26**, 211–252.

Brun V, Dupuist A, Adrait A, Marcellin M, Thomas D, Court M, Vandenesch F and Garin J 2007 Isotope-labeled protein standards. *Mol. Cell. Proteomics.* **6**, 2139–2149.

Cahill DJ and Nordhoff E 2003 Protein arrays and their role in proteomics. *Adv. Biochem. Eng. Biotechnol.* **83**, 177–187.

Callister SJ, Barry RC, Adkins JN, Johnson ET, Qian WJ, Webb-Robertson BJM, Smith RD and Lipton MS 2006 Normalization approaches for removing systematic biases associated with mass spectrometry and label-free proteomics. *J. Proteome Res.* **5**, 277–286.

Carbone A, Zinovyev A and Képès F 2003 Codon adaption index as a measure of dominating codon bias. *Bioinformatics* **19**, 2005–2015.

Carrillo B, Yanofsky C, Laboissiere S, Nadon R and Kearney RE 2010 Methods for combining peptide intensities to estimate relative protein abundance. *Bioinformatics* **26**, 98–103.

Carvalho PC, Hewel J, Barbosa VC and Yates III, JR. 2009 Identifying differences in protein expression levels by spectral counting and feature selection. *Genet. Mol. Res.* **7**, 342–356.

Chakraborty AB, Berger SJ and Gebler JC 2007 Use of an integrated ms - multiplexed ms/ms data acquisition strategy for high-coverage peptide mapping studies. *Rapid Commun. Mass. Spectrom.* **21**, 730–744.

Cham MJA, Bianco L and Bessant C 2010 Free computational resources for designing selected reaction monitoring transitions. *Proteomics* **10**, 1106–26.

Cheng F, Blackburn K, Lin Y, Goshe MB and Williamson JD 2009 Absolute protein quantification by lc/ms^c for global analysis of salicylic acid-induced plant protein secretion responses. *J. Proteome Res.* **8**, 82–93.

Chiogna M, Massa MS, Risso D and Romualdi C 2009 A comparison on effects of normalization in the detection of differentially expressed genes. *BMC Bioinformatics.* published online at www.biomedcentral.com/1471-2105/10/61.

Choi H, Fermin D and Nesvizhskii AI 2008 Significance analysis of spectral count data in label-free shotgun proteomics. *Mol. Cell Proteomics* **7**, 2373–2385.

Cleveland WS 1979 Robust locally weighted regression and smoothing scatterplots. *J. Am. Stat. Ass.* **74**, 829–836.

Colaert N, Gevaert K and Martens L 2011a Ribar and xribar: Methods for reproducible relative ms/ms-based label-free protein quantification. *J. Proteome Res.* **10**, 3183–3189.

Colaert N, Huele CV, Degroeve S, Staes A, Vandekerckhove J, Gevaert K and Martens L 2011b Combining quantitative proteomics data processing workflows for greater sensitivity. *Nat. Methods* **8**, 481–483.

Colaert N, Vandekerckhove J, Gevaert K and Martens L 2011c A comparison of ms2-based label-free quantitative proteomic techniques with regards to accuracy and precision. *Proteomics* **11**, 1110–1113.

Colinge J, Chiappe D, Lagache S, Moniatte M and Bougueleret L 2005 Differential proteomics via probabilistic peptide identification scores. *Anal. Chem.* **77**, 596–606.

Cook RD 1977 Detection of influential observations in linear regression. *Technometrics* **19**, 15–18.

Cooper B, Feng J and Garrett WM 2010 Relative, label-free protein quantization: Spectral counting error statistics from nine replicate mudpit samples. *J. Am. Soc. Mass Spectrom.* **21**, 1534–1546.

Côtè RG, Jones P, Martens L, Kerrien S, Reisinger F, Lin Q, Leinonen R, Apweiler R and Hermjakob H 2007 The protein identifier cross-referencing (picr) service: reconciling protein identifiers across multiple source databases. *BMC Bioinformatics.* Published online at www.biomedcentral.com/1471-2105/8/401.

Cox J and Mann M 2008 Maxquant enables high peptide identification rates, individualized p.p.b.-range mass accuracies and proteome-wide protein quantification. *Nat. Biotechnol.* **26**, 1367–1372.

Cox J and Mann M 2009 Computational principles of determining and improving mass precision and accuracy for proteome measurements in an orbitrap. *J. Am. Soc. Mass Spectrom.* **20**, 1477–1485.

Cox J, Matic I, Hilger M, Nagaraj N, Selbach M, Olsen JV and Mann M 2009 A practical guide to the maxquant computational platform for silac-based quantitative proteomics. *Nat. Protoc.* **4**, 698–705.

Cox J, Neuhauser N, Michalski A, Scheltema RA, Olsen JV and Mann M 2011 Andromeda: A peptide search engine integrated into the maxquant environment. *J. Proteome Res.* **10**, 1794–1805.

Craig R, Cortens JC, Fenyo D and Beavis RC 2006 Using annotated peptide mass spectrum libraries for protein identification. *J. Proteome Res.* **5**, 1843–1849.

Craig R, Cortens JP and Beavis RC 2004 Open source system for analyzing, validating, and storing protein identification data. *J. Proteome Res.* **3**, 1234–1242.

Craig R, Cortens JP and Beavis RC 2005 The use of proteotypic peptide libraries for protein identification. *Rapid Commun. Mass Spectrom.* **19**, 1844–1850.

Currie LA 1968 Limits for qualitative detection and quantitative determination. application to radiochemistry. *Anal. Chem.* **40**, 586–593.

Dayon L, Hainard A, Licker V, Turck N, Kuhn K, Hochstrasser D, Burkhard P and Sanchez J 2008 Relative quantification of proteins in human cerebrospinal fluids by ms/ms using 6-plex isobaric tags. *Anal. Chem.* **80**, 2923–2931.

Degroeve S, Colaert N, Vandekerckhove J, Gevaert K and Martens L 2011 A reproducibility-based evaluation procedure for quantifying the differences between ms/ms peak intensity normalization methods. *Proteomics* **11**, 1172–1180.

Dehaene S, Izard V, Spelke E and Pica P 2008 Log or linear? distinct intuitions of the number scale in western and amazonian indigene cultures. *Science* **320**, 5880.

Deininger S, Cornett DS, Paape R, Becker M, Pineau C, Rauser S, Walch A and Wolski E 2011 Normalization in maldi-tof imaging datasets of proteins: practical considerations. *Anal. Bioanal. Chem.* **401**, 167–181.

Desiere F, Deutsch EW, King NL, Nesvizhskii AI, Mallick P, Eng J, Chen S, Eddes J, Loevenich SN and Aebersold R 2006 The peptideatlas project. *Nucleic Acids Res.* **34**, 655–658.

DeSouza LV, Taylor AM, Li W, Minkoff MS, Romaschin AD, Colgan TJ and Siu KWM 2008 Multiple reaction monitoring of mtraq-labeled peptides enables absolute quantification of endogenous levels of a potential cancer marker in cancerous and normal endometrial tissues. *J. Proteome Res.* **7**, 3525–3534.

Deutsch EW, Chambers M, Neumann S, Levander F, Binz PA, Shofstahl J, Campbell DS, Mendoza L, Ovelleiro D, Helsens K, Martens L, Aebersold R, Moritz RL and Brusniak MY 2012 Traml–a standard format for exchange of selected reaction monitoring transition lists. *Mol. Cell. Proteomics* **11**(4), R111.01504.

Dicker L, Lin X and Ivanov AR 2010 Increased power for the analysis of label-free lc-ms/ms proteomics data by combining spectral counts and peptide peak attributes. *Mol. Cell Proteomics* **9**, 2704–2718.

Dixon WJ 1953 Processing data for outliers. *Biometrics* **9**, 74–89.

Domanski D, Murphy LC and Borchers CH 2010 Assay development for the determination of phosphorylation stoichiometry using multiple reaction monitoring methods with and without phosphatase treatment: application to breast cancer signaling pathways. *Anal. Chem.* **82**, 5610–5620.

Domon B and Aebersold R 2010 Options and considerations when selecting a quantitative proteomics strategy. *Nat. Biotechnol.* **28**, 710–721.

Dost B, Bandeira N, Li X, Shen Z, Briggs S and Bafna V 2009 Shared peptides in mass spectrometry based protein quantification *Research in Computational Molecular Biology Lecture Notes in Computer Science*, vol. 5541, pp. 356–371. Springer Link.

Dudoit S, Yang YH, Callow MJ and Speed TP 2002 Statistical methods for identifying differentially expressed genes in replicated cdna microarray experiments. *Stat. Sin.* **12**, 111–139.

Durbin BP, Hardin JS, Hawkins DM and Rocke DM 2002 A variance-stabilization transformation for gene-expression microarray data. *Bioinformatics* **18**, S105–S110.

Eidhammer I, Flikka K, Martens L and Mikalsen SO 2007 *Computational Methods for Mass Spectrometry Proteomics.* John Wiley & Sons, Ltd.

Eissler CL, Bremmer SC, Martinez JS, Parker LL, Charbonneau H and Hall MC 2011 A general strategy for studying multisite protein phosphorylation using label-free selected reaction monitoring mass spectrometry. *Anal. Biochem.* **418**, 267–275.

Ernoult E, Gamelin E and Guette C 2008 Improved proteome coverage by using itraq labelling and peptide offgel fractionation. *Proteome Sci.* Published online at www.proteomesci.com/content/6/1/27.

Farell PJ and Roger-Stewart K 2006 Comprehensive study of tests for normality and symmetry: extending the spiegelhalter test. *J. Stat. Comp. Sim.* **76**, 803–816.

Fox JM and Erill I 2010 Relative codon adaptation: A generic codon bias index for prediction of gene expression. *DNA Res.* **17**, 185–196.

Fu N, Drinnenberg I, Kelso J, Wu JR, Pääbo S, Zeng R and Khaitovich P 2007 Comparison of protein and mrna expression evolution in human and chimpanzees. *PLoS One* **2**, e216.

Gan CS, Chong PK, Pham TK and Wright PC 2007 Technical, experimental, and biological variations in isobaric tags for relative and absolute quantitation (itraq). *J Proteome Res.* **6**, 821–827.

Garcia E and Puimedón J 2004 Optimal integration window for peak area estimation. *J. Anal. At. Spectrom* **19**, 1391–1393.

Geiger T, Cox J, Ostasiewicz P, Wisniewski JR and Mann M 2010 Super-silac mix for quantitative proteomics of human tumor tissue. *Nat. Method* **7**, 383–387.

Geiger T, Wisniewski JR, Cox J, Zanivan S, Kruger M, Ishihama Y and Mann M 2011 Use of stable isotope labeling by amino acids in cell culture as a spike-in standard in quantitative proteomics. *Nat. Protoc.* **6**, 147–157.

Gerber SA, Rush J, Stemman O, Kirschner MW and Gygi SP 2003 Absolute quantification of proteins and phosphoproteins from cell lysates by tandem ms. *Proc. Natl. Acad. Sci. USA* **100**, 6940–6945.

Greenbaum D, Colangelo C, Williams K and Gerstain M 2003 Comparing protein abundance and mrna expression levels on a genomic scale. *Genom. Biol.* **4**, 117.

Griffin NM, Yu J, Long F, Oh P, Shore S, Li Y, Koziol JA and Schnitzer JE 2010 Label-free, normalized quantification of complex mass spectrometry data for proteomic analysis. *Nat. Biotechnol.* **28**, 83–91.

Griffin TJ, Gygi SP, Ideker T, Rist B, Eng J, Hood L and Aebersold R 2002 Complementary profiling of gene expression at the transcriptome and proteome levels in saccharomyces cerevisiae. *Mol. Cell. Proteomics* **1**, 323–333.

Grubbs FE 1969 Procedures for detecting outlying observations in samples. *Technometrics* **11**, 1–21.

Guo Y, Xiao P, Lei S, Deng F, Xiao GG, Liu Y, Chen X, Li L, Wu S, Chen Y, Jiang H, Tan L, Xie J, Zhu X, Liang S and Deng H 2008 How is mrna expression predictive for protein expression? a correlation study on human circulating monocytes. *Acta Biochim. Biophys. Sin.* **40**, 426–436.

Gygi SP, Rist B, Gerber SA, Turecek F, Gelb MH and Aebersold R 1999 Quantitative analysis of complex protein mixtures using isotope-coded affinity tags. *Nat. Biotechnol.* **17**, 994–999.

Hanke S, Besir H, Oesterhelt D and Mann M 2008 Absolute silac for accurate quantitation of proteins in complex mixtures down to the attomole level. *J. Proteome Res.* **7**, 1118–1130.

Hegeman AD, Harms AC, Sussman MR, Bunner AE and Harper JF 2004 An isotope labeling strategy for quantifying the degree of phosphorylation at multiple sites in proteins. *J. Am. Soc. Mass Spectrom.* **15**, 647–653.

Hermjakob H *et al.* 2004 The hupo psi's molecular interaction format – a community standard for the representation of protein interaction data. *Nat. Biotechnol.* **22**, 177–183.

Higgs RE, Knierman MD, Gelfanova V, Butler JP and Hale JE 2005 Comprehensive label-free method for the relative quantification of proteins from biological samples. *J. Proteome Res.* **4**, 1442–1450.

Hill EG, Schwacke JH, Comte-Walters S, Slate EH, Oberg AL, Eckel-Passow JE, Therneau T and Schey K 2008 A statistical model for itraq data analysis. *J. Proteome Res.* **7**, 3091–3101.

Hilterbrand A, Saelens J and Putonti C 2012 Cbdb: The codon bias database. *BMC Bioinformatics*.

Hsu JL, Huang SY and Chen SH 2006 Dimethyl multiplexed labeling combined with microcolumn separation and ms analysis for time course study in proteomics. *Electrophoresis* **27**, 3652–3660.

Hsu JL, Huang SY, Chow NH and Chen SH 2003 Stable-isotope dimethyl labeling for quantitative proteomics. *Anal. Chem.* **75**, 6843–6852.

Huber W, A. von Heydebreck, Sültmann H, Poustka A and Vingron M 2002 Variance stabilization applied to microarray data calibration and to the quantification of differential expression. *Bioinformatics* **18**, S96–S104.

Ishiama Y, Oda Y, Tabata T, Sato T, Nagasu T, Rappsilber J and Mann M 2005a Exponentially modified protein abundance index (empai) for estimation of absolute protein amount in proteomics by the number of sequenced peptides per protein. *Mol. Cell. Proteomics* **4**, 1265–1272.

Ishiama Y, Sato T, Tabata T, Miyamoto N, Sagane K, Nagasu T and Oda Y 2005b Quantitative mouse brain proteomics using culture-derived isotope tags as internal standards. *Nat. Biotechnol.* **23**, 617–621.

Issaq HJ, Waybright TJ and Veenstra TD 2011 Cancer biomarker discovery: Opportunities and pitfalls in analytical methods. *Electrophoresis* **32**, 967–975.

Jansen R, Bussemaker HJ and Gerstein M 2003 Revisiting the codon adaption index from a whole-genome perspective: analyzing the relationship between gene expression and codon occurrence in yeast using a variety of models. *Nucleic Acids Res.* **31**, 2242–2251.

Jin LL, Tong J, Prakash A, Peterman SM, St-Germain JR, Taylor P, Trudel S and Moran MF 2010 Measurement of protein phosphorylation stoichiometry by selected reaction monitoring mass spectrometry. *J. Proteome Res.* **9**, 2752–2761.

Jin S, Daly DS, Springer DL and Miller JH 2008 The effects of shared peptides on protein quantitation in label-free proteomics by lc/ms/ms. *J. Proteome Res.* **7**, 164–169.

Jones AR, Eisenacher M, Mayer G, Kohlbacher O, Siepen J, Hubbard S, Selley J, Searle B, Shofstahl J, Seymour S, Julian R, Binz PA, Deutsch EW, Hermjakob H, Reisinger F, Griss J, Vizcaino JA, Chambers M, Pizarro A and Creasy D 2012 The mzidentml data standard for mass spectrometry-based proteomics results. *Mol. Cell. Proteomics*. Ahead of print.

Käll L, Storey JD and Noble WS 2009 Qvality: non-parametric estimation of q-values and posterior error probabilities. *Bioinformatics* **25**, 964–966.

Käll L, Storey JD, MacCoss MJ and Noble WS 2008 Assigning significance to peptides identified by tandel mass spectrometry using decoy databases. *J. Proteome Res.* **7**, 29–34.

Karp NA, Huber W, Sadowski PG, Charles PD, Hester SV and Lilley KS 2010 Addressing accuracy and precision issues in itraq quantitation. *Mol. Cell. Proteomics* **9**, 1885–1897.

Karpievitch Y, Stanley J, Taverner T, Huang J, Adkins JN, Ansong C, Heffron F, Metz TO, Qian WJ, Yoon H, Smith RD and Dabney AR 2009a A statistical framework for protein quantitation in bottom-up ms-based proteomics. *Bioinformatics* **25**, 2028–2034.

Karpievitch Y, Taverner T, Adkins JN, Callister SJ, Anderson GA, Smith RD and Dabney AR 2009b Normalization of peak intensities in bottom-up ms-based proteomics using singular value decomposition. *Bioinformatics* **25**, 2573–2580.

Keshamouni VG, Jagtap P, Michailidis G, Strahler JR, Kuick R, Reka AK, Papoulias P, Krishnapuram R, Srirangam A, Standiford TJ, Andrews PC and Omenn GS 2009 Temporal quantitative proteomics by itraq 2d-lc-ms/ms and corresponding mrna expression analysis identify post-transcriptional modulation of actin-cytoskeleton regulators during tgf-β-induced epithelial-mesenchymal transition. *J. Proteome Res.* **8**, 35–47.

Keshamouni VG, Michailidis G, Grasso CS, Anthwal S, Strahler JR, Walker A, Arenberg DA, Reddy RC, Akulapalli S, Thannickal VJ, Standiford TJ, Andrews PC and Omenn GS 2007 Differential protein expression profiling by itraq-2dlc-ms/ms of lung cancer cells undergoing epithelial-mesenchymal transition reveals a migratory/invasive phenotype. *J. Proteome Res.* **5**, 1143–1154.

Keshishian H, Addona T, Burgess M, Kuhn E and Carr SA 2007 Quantitative, multiplexed assays for low abundance proteins in plasma by targeted mass spectrometry and stable isotope dilution. *Mol. Cell. Proteomics* **6**, 2212–2229.

Keshishian H, Addona T, Burgess M, Mani DR, Shi X, Kuhn E, Sabatine MS, Gerszten RE and Carr SA 2009 Quantification of cardiovascular biomarkers in patient plasma by targeted mass spectrometry and stable isotope dilution. *Mol. Cell. Proteomics* **8**, 2339–2349.

Kettenbach AN, Rush J and Gerber SA 2011 Absolute qunatification of protein and post-translational modification abundance with stable isotope-labeled synthetic peptides. *Nat. Protoc.* **6**, 175–186.

Kijanka G and Murphy D 2009 Protein arrays as tools for serum autoantibody marker discovery in cancer. *J. Proteomics* **20**, 936–944.

Kirkpatrick DS, Gerber SA and Gygi SP 2005 The absolute quantification strategy: a general procedure for the qunatification of proteins and post-translational modifications. *Methods* **35**, 265–273.

Kito K, Ota K, Fujita T and Ito T 2007 A synthetic protein approach toward accurate mass spectrometric quantification of component stoichiometry of multiprotein complexes. *J. Proteome Res.* **6**, 792–800.

Klecka WR 1980 *Discriminant Analysis*. SAGE Publications.

Koehler CJ, Strozynski M, Kozielski F, Treumann A and Thiede B 2009 Isobaric peptide termini labeling for ms/ms-based quantitative proteomics. *J. Proteome Res.* **8**, 4333–4341.

Kultima K, Nilsson A, Scholz B, Rossbach UL, Fälth M and André PE 2009 Development and evaluation of normalization methods for label-free relative quantification of endogenous peptides. *Mol. Cell. Proteomics* **8**, 2285–2295.

Kyte J and Doolittle RF 1982 A simple method for displaying the hydropathic character of a protein. *J. Mol. Biol.* **157**, 105–132.

LaBaer J 2005 So, you want to look for biomarkers (introduction to the special biomarkers issue). *J. Proteome Res.* **4**, 1053–1059.

Lacerda CMR, Xin L, Rogers I and Reardon KF 2008 Analysis of itraq data using mascot and peaks quantification algorithms. *Brief. Funct. Genomic Proteomic* **7**, 119–126.

Lam H, Deutsch EW, Eddes JS, Eng JK, King N, Stein SE and Aebersold R 2007 Development and validation of a spectral library searching method for peptide identification from ms/ms. *Proteomics* **7**, 655–667.

Lange E, Grpl C, Schulz-Trieglaff O, Leinenbach A, Huber C and Reinert K 2007 A geometric approach for the alignment of liquid chromatography mass spectrometry data. *Bioinformatics* **23**, i273–i281.

Lange V, Picotti P, Domon B and Aebersold R 2008 Selected reaction monitoring for quantitative proteomics: a tutorial. *Mol. Syst. Biol.* Issue 222.

Li F, Nie L, Wu G, Qiao J and Zhang W 2011 Prediction and characterization of missing proteomic data in desulfovibrio vulgaris. *Comp. Funct. Genomics*. Published online doi: 10.1155/2011/780973.

Li X, Yi EC, Kemp CJ, Zhang H and Aebersold R 2005 A software suite for the generation and comparison of peptide arrays from sets of data collected by liquid chromatography-mass spectrometry. *Mol. Cell. Proteomics* **4**, 1328–1340.

Li X, Zhang H, Ranish JA and Aebersold R 2003 Automated statistical analysis of protein abundance ratios from data generated by stable-isotope dilution and tandem mass spectrometry. *Anal. Chem.* **75**, 6648–6657.

Li YF, Arnold RJ, Li Y, Radivojac P, Sheng Q and Tang H 2009 A bayesian approach to protein inference problem in shotgun proteomics. *J. Comput. Biol.* **16**, 1183–1193.

Lin WT, Hung WN, Yian YH, Wu KP, Han CL, Chen YR, Chen YJ, Sung TY and Hsu WL 2006 Multi-q: a fully automated tool for multiplexed protein quantitation. *J. Proteome Res.* **5**, 2328–2338.

Linnet K and Kondratovich M 2004 Partly nonparametric approach for determining the limit of detection. *Clin. Chem.* **50**, 732–740.

Liu H, Sadygov RG and Yates III, JR 2004 A model for random sampling and estimation of relative protein abundance in shotgun proteomics. *Anal. Chem.* **76**, 4193–4201.

Liu Z, Cao J, He Y, Qiao L, Xu C, Lu H and Yang P 2010 Tandem ^{18}o stable isotope labeling for quantification of n-glycoproteome. *J. Proteome Res.* **9**, 227–236.

Lu P, Vogel C, Wang R, Yao X and Marcotte EM 2007 Absolute protein expression profiling estimates the relative contributions of transcripional and translational regulation. *Nat. Biotechnol.* **25**, 117–124.

Maercker C 2005 Protein arrays in functional genome research. *Biosci. Rep.* **25**, 57–70.

Maier T, Güell M and Serrano L 2009 Correlation of mrna and protein in complex biological samples. *FEBS Lett.* **583**, 3966–3973.

Maier T, Schmidt A, Güell M, Kühner S, Gavin AC, Aebersold R and Serrano L 2011 Quantification of mrna and protein and integration with protein turnover in a bacterium. *Mol. Syst. Biol.* Published online doi:10.1038/msb.2011.38.

Malmström J, Beck M, Schmidt A, Lange V, Deutch EW and Aebersold R 2009 Proteome-wide cellular protein concentrations of the human pathogen leptospira interrogans. *Nature* **460**, 762–766.

Martens L, Chambers M, Sturm M, Kessner D, Levander F, Shofstahl J, Tang WH, Römpp A, Neumann S, Pizarro AD, Montecchi-Palazzi L, Tasman N, Coleman M, Reisinger F, Souda P, Hermjakob H, Binz PA and Deutsch EW 2011 mzml–a community standard for mass spectrometry data. *Mol. Cell. Proteomics* **10**(1), R110.000133.

Martens L, Hermjakob H, Jones P, Adamski M, Taylor C, States D, Gevaert K, Vandekerckhove J and Apweiler R 2005 Pride: the proteomics identifications database. *Proteomics* **5**, 3537–3545.

Montecchi-Palazzi L, Kerrien S, Reisinger F, Aranda B, Jones AR, Martens L and Hermjakob H 2009 The psi semantic validator: a framework to check miape compliance of proteomics data. *Proteomics* **9**, 5112–5119.

Motulsky H 1995 *Intuitive Biostatistics*. Oxford University Press.

Mueller LN, Brusniak MI, Mani DR and Aebersold R 2008 An assessment of software solutions for the analysis of mass spectrometry based quantitative proteomics data. *J. Proteome Res.* **7**, 51–61.

Mueller LN, Rinner O, Schmidt A, Letarte S, Bodenmiller B, Brusniak MY, Vitek O, Aebersold R and Möller M 2007 Superhirn a novel tool for high resolution lc-ms-based peptide/protein profiling. *Proteomics* **7**, 3470–3480.

Muth T, Keller D, Puetz SM, Martens L, Sickmann A and Boehm AM 2010 jtraqx: a free, platform independent tool for isobaric tag quantitation at the protein level. *Proteomics* **10**, 1223–1228.

Na S and Paek E 2006 Quality assessment of tandem mass spectra based on cumulative intensity normalization. *J. Proteome Res.* **5**, 3241–3248.

Neilson KA, Ali NA, Muralidharan S, Mirzaei M, Mariani M, Assadourian G, Lee A, van Sluyter SC and Haynes PA 2011 Less label, more free: Approaches in label-free quantitative mass spectrometry. *Proteomics* **11**, 1–19.

Nelson DL and Cox MM 2004 *Lehninger Principles of Biochemistry*. Worth Publishers.

Nesvizhskii AI and Aebersold R 2005 Interpretation of shotgun proteomic data. *Mol. Cell. Proteomics* **4**, 1419–1440.

Nie L, Wu G and Zhang W 2005 Correlation between mrna and protein abundance in deaulfovibrio vulgaris: A multiple regression to identify sources of variations. *Biochem. Biophys. Res. Commun.* **339**, 603–610.

Nie L, Wu G and Zhang W 2006 Correlation of mrna expression and protein abundance affected by multiple sequence features related to translational efficiency in desulfovibrio vulgaris: A quantitative analysis. *Genetics* **4**, 2229–2243.

Nilsson P, Paavilainen L, Larsson K, Odling J, Sundberg M, Andersson AC, Kampf C, Persson A, Al-Khaliliand SWC, Ottosson J, Björling E, Hober S, Wernérus H, Wester K, Pontén F and Uhlen M 2005 Towards a human proteome atlas: high-throughput generation of mono-specific antibodies for tissue profiling. *Proteomics* **5**, 4327–4337.

Norris J, Cornett DS, Mobley JA, Andersson M, Seeley EH, Chaurand P and Caprioli RM 2007 Processing maldi mass spectra to improve mass spectral direct tissue analysis. *Int. J. Mass Spectrom.* **260**, 212–221.

Norris M and Siegfried DR 2011 *Anatomy and Physiology For Dummies, 2nd Edition*. Wiley.

Oberg AL, Mahoney DW, Eckel-Passow JE, Malone CJ, Wolfinger RD, Hill EG, Cooper LT, Onuma OK, Spiro C, Therneau TM and H. Robert Bergen, III 2008 Statistical analysis of relative labeled mass spectrometry data from complex samples using anova. *J. Proteome Res.* **7**, 225–233.

Old WM, Meyer-Arendt K, Aveline-Wolf L, Piercet KG, Mendozat A, Sevinsky JR, Resing KA and Ahn NG 2005 Comparison of label-free methods for quantifying human proteins by shotgun proteomics. *Mol Cell. Proteomics* **4**, 1487–1502.

Ong S and Mann M 2005 Mass spectrometry-based proteomics turns quantitative. *Nat. Chem. Biol.* **1**, 252–262.

Panchaud A, Affolter M, Moreillon P and Kussmann M 2008 Experimental and computational approaches to quantitative proteomics: Status quo and outlook. *J. Proteomics* **71**, 19–33.

Park P, Kohane IS and Kim JH 2003a Rank-based nonlinear normalization of oligonucleotide arrays. *Gen. Inf.* **1**, 94–100.

Park T, Yi SG, Kang SH, Lee SY, Lee YS and Simon R 2003b Evaluation of normalization methods for microarray data. *BMC Bioinformatics*. Published online at www.biomedcentral.com/1471-2105/4/33.

Pascal LE, True LD, Campbell DS, Deutsch EW, Risk M, Coleman IM, Eichner LJ, Nelson PS and Liu AY 2008 Correlation of mrna and protein levels; cell type-specific gene expression of cluster designation antigens in the prostate. *BMC Genomics*. Published online at www.biomedcentral.com/1471-2164/9/246.

Pavelka N, Fournier ML, Swanson SK, Pelizzola M, Ricciardi-Castagnoli P, Florens L and Washburn MP 2008 Statistical similarities between transcriptomics and quantitative shotgun proteomics data. *Mol. Cell. Proteomics* **7**, 631–644.

Pearson K 1895 Contributions to the mathematical theory of evolution, ii: Skew variation in homogeneous material. *Philosophical Transactions of the Royal Society of London* **186**, 343–414.

Pedreschi R, Hertog ML, Carpentier SC, Lammertyn J, Robben J, Noben JP, Panis B, Swennen R and Nicola BM 2008 Treatment of missing values for multivariate statistical analysis of gel-based proteomics data. *Proteomics* **8**, 1371–1383.

Petrakis L 1967 Spectral line shapes. *J. Chem. Educ.* **44**, 432–436.

Phanstiel D, Zhang Y, Marto JA and Coon JJ 2008 Peptide and protein quantification using itraq with electron transfer dissociation. *J. Am. Soc. Mass. Spectrom.* **19**, 1255–1262.

Pichler P, Köcher T, Holzmann J, Mazanek M, Taus T, Ammerer G and Mechtler K 2010 Peptide labeling with isobaric tags yields higher identification rates using itraq 4-plex compared to tmt 6-plex and itraq 8-plex on ltq orbitrap. *Anal. Chem.* **82**, 6549–6558.

Podwojski K, Fritsch A, Chamrad DC, Paul W, Sitek B, Mutzel P, Stephan C, Meyer HE, Urfer W, Ickstadt K and Rahnenführer J 2009 Retention time alignment algorithms for lc/ms data must consider nonlinear shifts. *Bioinformatics* **25**, 758–764.

Prince JT and Marcotte EW 2006 Chromatographic alignment of esi-lc-ms proteomics data sets by ordered bijective interpolated warping. *Anal. Chem.* **78**, 6140–6152.

Rabilloud T, Chevallet M, Luche S and Lelong C 2010 Two-dimensional gel electrophoresis in proteomics: Past, present and future. *J. Proteomics* **73**, 2064–2077.

Rappsilber J, Ryder U, Lamond AI and Mann M 2002 Large-scale proteomic analysis of the human spliceosome. *Genome Res.* **12**, 1231–1245.

Reid JD, Parker CE and Borchers CH 2007 Protein arrays for biomarker discovery. *Curr. Opin. Mol. Ther.* **9**, 216–221.

Reidegeld KA, Eisenacher M, Kohl M, Chamrad D, Körting G, Blüggel M, Meyer HE and Stephan C 2008 An easy-to-use decoy database builder software tool, implementing different decoy strategies for false discovery rate calculation in automated ms/ms protein identifications. *Proteomics* **8**, 1129–1137.

Reiter L, Rinner O, Picotti P, Hüttenhain R, Beck M, Brusniak MY, Hengartner MO and Aebersold R 2011 mprophet: automated data processing and statistical validation for large-scale srm experiments. *Nat. Methods* **8**, 430–435.

Reynolds KJ, Yao X and Fenselau C 2002 Proteolytic ^{18}o labeling for comparative proteomics: Evaluation of endoprotease glu-c as the catalytic agent. *J. Proteome Res.* **1**, 27–33.

Rorabacher DB 1991 Statistical treatment for rejection of deviant values: critical values of dixon's q parameter and related subrange ratios at the 95% confidence level. *Anal. Chem.* **63**, 139–146.

Rosner B 1983 Percentage points for a generalized esd many-outlier procedure. *Technometrics* **25**(2), 165–172.

Ross PL, Huang YN, Marchese JN, Williamson B, Parker K, Hattan S, Khainovski N, Pillai S, Dey S, Daniels S, Purkayastha S, Juhasz P, Martin S, Bartlet-Jones M, He F, Jacobson A and Pappin DJ 2004 Multiplexed protein quantitation in saccharomyces cerevisiae using amine-reactive isobaric tagging reagents. *Mol. Cell. Proteomics* **3**, 1154–1169.

Roy 2001 *amc technical brief*. Analytical Methods Committee.

Roy P, Truntzer C, Maucort-Boulch D, Jouve T and Molinari N 2011 Protein mass spectra data analysis for clinical biomarker discovery: a global review. *Brief. Bioinform.* **12**, 176–186.

Royston JP 1982 An extension to shapiro and wilk's *w* test for normality to large samples. *Appl. Stat.* **31**, 115–124.

Schmidt A, Kellermann J and Lottspeich F 2005 A novel strategy for quantitative proteomics using isotope-coded protein labels. *Proteomics* **5**, 4–15.

Schwanhäusser B, Busse D, Li N, Dittmar G, Schuchhardt J, Wolf J, Chen W and Selbach M 2011 Global quantification of mammalian gene expression control. *Nature* **473**, 337–342.

Scigelova M and Makarov A 2006 Orbitrap mass analyzer - overview and applications in proteomics. *Proteomics, Supplement practical proteomics* **6**, 16–21. Issue 2.

Sharp PM and Li WH 1987 The codon adaptation index–a measure of directional synonymous codon usage bias, and its potential applications. *Nucleic Acids Res.* **15**, 1281–1295.

Silva JC, Denny R, Dorschel CA, Gorenstein M, Kass IJ, Li GZ, McKenna T, Nold MJ, Richardson K, Young P and Geromanos S 2005 Quantitative proteomic analysis by accurate mass retention time pairs. *Anal. Chem.* **77**, 2187–2200.

Silva JC, Gorenstein MV, Li GZ, Vissers JPC and Geromanos S 2006 Absolute quantification of proteins by lcms. *Mol. Cel. Proteomics* **5**, 144–156.

Singh S, Springer M, Steen J, Kirschner MW and Steen H 2009 Flexiquant: a novel tool for the absolute quantification of proteins, and the simultaneous identification and quantification of potentially modified peptides. *J. Proteome Res.* **8**, 2201–2210.

Smith CA, Want EJ, O'Maille G, Abagyan R and Siuzdak G 2006 Xcms: processing mass spectrometry data for matabolite profiling using nonlinear peak alignment, matching, and identification. *Anal. Chem.* **78**, 779–787.

Song X, Bandow J, Sherman J, Baker JD, Brown PW, McDowell MT and Molloy MP 2008 itraq experimental design for plasma biomarker discovery. *J. Proteome Res.* **7**, 2952–2958.

Staes A, Demol H, Damme JV, Martens L, Vandekerckhove J and Gevaert K 2004 Global differential non-gel proteomics by quantitative and stable labeling of tryptic peptides with oxygen-18. *J. Proteome Res.* **3**, 786–91.

Steen H, Jebanathirajah JA, Springer M and Kirschner MW 2005 Stable isotope-free relative and absolute quantitation of protein phosphorylation stoichiometry by ms. *Proc. Natl. Acad. Sci.* **102**, 3948–3953.

Stevens SS 1946 On the theory of scales of measurement. *Science* **103**, 677–680.

Storey JD and Tibshirani R 2003 Statistical significance for genomewide studies. *Proc. Natl. Acad. Sci.* **100**, 9449–9445.

Tambor V, Fucíková A, Lenco J, Kacerovský M, Rehácek V, Stulk J and Pudil R 2009 Application of proteomics in biomarker discovery: a primer for the clinician. *Physiol. Res.* **59**, 471–497.

Tautenhahn R, Patti GJ, Kalisiak E, Miyamoto T, Schmidt M, Lo FY, McBee J, Baliga NS and Siuzdak G 2010 metaxcms: Second-order analysis of untargeted metabolomics data. *Anal. Chem.* **83**, 696–700.

Theodoridis S and Koutroumbas K 2006 *PATTERN RECOGNITION*. Academic Press.

Thompson A, Schäfer J, Kuhn K, Kienle S, Schwarz J, Schmidt G, Neumann T and Hamon C 2003 Tandem mass tags: A novel quantification strategy for comparative analysis of complex protein mixtures by ms/ms. *Anal. Chem.* **75**, 1895–1904.

Thompson B 1984 *Canonical Correlation Analysis*. SAGE Publications.

Tian Q, Stepaniants SB, Mao M, Weng L, Feetham MC, Doyle MJ, Yi EC, Dai H, Thorsson V, Eng J, Goodlett D, Berger JP, Gunter B, Linseley PS, Stoughton RB, Aebersold R, Collins SJ, Hanion WA and Hood LE 2004 Integrating genomic and proteomic analyses of gene expression in mammalian cells. *Mol. Cell. Proteomics* **3**, 960–968.

Tusher VG, Tibshirani R and Chu G 2001 Significance analysis of microarrays applied to the ionizing radiation response. *Proc. Natl. Acad. Sci.* **98**, 5116–5121.

Unwin RD, Griffiths JR and Whetton AD 2010 Simultaneous analysis of relative protein expression levels across multiple samples using itraq isobaric tags with 2d nano lc-ms/ms. *Nat. Protoc.* **5**, 1574–1581.

van Iterson M, Boer JM and Menezes RX 2010 Filtering, fdr and power. *BMC Bioinformatics* **11**, 450.

Vandenbogaert M, Li-Thiao-T1 S, Kaltenbach HM, Zhang R, Aittokallio T and Schwikowski B 2008 Alignment of lc-ms images, with applications to biomarker discovery and protein identification. *Proteomics* **8**, 650–672.

Vaudel M, Sickmann A and Martens L 2010 Peptide and protein quantification: A map of the minefield. *Proteomics* **10**, 650–670.

Velleman PF and Wilkinson L 1993 Nominal, ordinal, interval, and ratio typologies are misleading. *The Am. Stat.* **47**(1), 65–72.

Verma SP and Quiroz-Ruiz A 2006a Critical values for 22 discordancy test variants for outliers in normal samples up to sizes 100, and application in science and engineering. *Revista Mexicana de Ciencias Geológicas* **23**, 302–319.

Verma SP and Quiroz-Ruiz A 2006b Critical values for for six dixon tests for outliers in normal samples up to sizes 100, and application in science and engineering. *Revista Mexicana de Ciencias Geológicas* **23**, 133–161.

Vizcano JA, Martens L, Hermjakob H, Julian RK and Paton NW 2007 The psi formal document process and its implementation on the psi website. *Proteomics* **7**, 2355–2357.

Vogel C, de Sousa Abreu, R., Ko D, Le SY, Shapiro BA, Burns SC, Sandhu D, Boutz DR, Marcotte EM and Penalva LO 2010 Sequence signatures and mrna concentration can explain two-thirds of protein abundance variation in a human cell line. *Mol. Syst. Biol.* **6**, 400.

von Hippel PR 2005 Mean, median, and skew: Correcting a textbook rule. *J. Stat. Educ.* Published online www.amstat.org/publications/jse/v13n2/vonhippel.html.

Voss B, Hanselmann M, Renard BY, Lindner MS, Köthe U, Kirchner M and Hamprecht FA 2011 Sima: Simultaneous multiple alignment of lc/ms peak lists. *Bioinformatics* **27**, 987–993.

Wang M, You J, Bemis KG, Tegeler TJ and Brown DPG 2008 Label-free mass spectrometry-based protein quantification technologies in proteomic analysis. *Brief. Funct. Genomic Proteomic* **7**, 329–339.

Wilson B, Liotta LA and Petricoin E 2010 Monitoring proteins and protein networks using reverse phase protein arrays. *Dis. Markers* **28**, 225–232.

Yao X, Freas A, Ramirez J, Demirev PA and Fenselau C 2001 Proteolytic 18o labeling for comparative proteomics: model studies with two serotypes of adenovirus. *Anal. Chem.* **73**, 2836–2842.

Yona G, Dirks W, Rahman S and Lin DM 2006 Effective similarity measures for expression profiles. *Bioinformatics* **22**, 1616–1622.

Yu EZ, Burba AEC and Gerstein M 2007 Pare: A tool for comparing protein abundance and mrna expression data. *BMC Bioinformatics* **8**, 309.

Zangar RC, Daly DS and White AM 2006 Elisa microarray technology as a high-throughput system for cancer biomarker validation. *Expert Rev. Proteomics* **3**, 37–44.

Zangar RC, Varnum SM, Covington CY and Smith RD 2004 A rational approach for discovering and validating cancer markers in very small samples using mass spectrometry and elisa microarrays. *Dis. Markers* **20**, 135–148.

Zee BM, Young NL and Garcia BA 2011 Quantitative proteomic approaches to studying histone modifications. *Curr. Chem. Genomics* **5**, 106–114.

Zeiler M, Straube1 WL, Lundberg E, Uhlen M and Mann M 2011 A protein epitope signature tag (prest) library allows silac-based absolute quantification and multiplexed determination of protein copy numbers in cell lines. *Mol. Cell Proteomics* **11**(3), O111.009613.

Zhang B, VerBerkmoes NC, Langston MA, Uberbacher E, Hettich RL and Samatova NF 2006 Detecting differential and correlated protein expression in label-free shotgun proteomics. *J. Proteome Res.* **5**, 2909–2918.

Zhang G and Neubert TA 2006 Automated comparative proteomics based on multiplex tandem mass spectrometry and stable isotope labeling. *Mol. Cell. Proteomics* **5**, 401–411.

Zhang H, Li XJ, Martin DB and Aebersold R 2003 Identification and quantification of n-linked glycoproteins using hydrazide chemistry, stable isotope labeling and mass spectrometry. *Nat. Biotechnol.* **21**, 660–666.

Zhang Y, Wen Z, Washburn MP and Florens L 2009 Effect of dynamic exclusion duration on spectral count based quantitative proteomics. *Anal. Chem.* **81**, 6317–6326.

Zhang Y, Wen Z, Washburn MP and Florens L 2010 Refinements to label free proteome quantitation: How to deal with peptides shared by multiple proteins. *Anal. Chem.* **82**, 2272–2281.

Zhou JY, Schepmoes AA, Zhang X, Moore RJ, Monroe ME, Lee JH, Camp, II, DG., Smith RD and Qian WJ 2010 Improved lcms/ms spectral counting statistics by recovering low-scoring spectra matched to confidently identified peptide sequences. *J. Proteome Res.* **9**, 5698–5704.

Zhou L, Beuerman RW, Chew AP, Koh SK, Cafaro TA, Urrets-Zavalia EA, Urrets-Zavalia JA, Li SFY and Serra HM 2009 Quantitative analysis of n-linked glycoproteins in tear fluid of climatic droplet keratopathy by glycopeptide capture and itraq. *J. Proteome Res.* **8**, 1992–2003.

Zybailov B, Mosley AL, Sardiu ME, Coleman MK, Florens L and Washburn MP 2006 Statistical analysis of membrane proteome expression changes ion saccharomyces cerevisiae. *J. Proteome Res.* **5**, 2339–2347.

Index

Computational and Statistical Methods for Protein Quantification by Mass Spectrometry,
First Edition. Ingvar Eidhammer, Harald Barsnes, Geir Egil Eide and Lennart Martens.
© 2013 John Wiley & Sons, Ltd. Published 2013 by John Wiley & Sons, Ltd.

Printed and bound by CPI Group (UK) Ltd, Croydon, CR0 4YY

27/10/2024

14580190-0003